Assembly Programming and the 8086 Microprocessor

Assembly Programming and the 8086 Microprocessor

D. S. JONES

Department of Mathematical Sciences, University of Dundee

Oxford New York Tokyo

OXFORD UNIVERSITY PRESS

1988

Oxford University Press, Walton Street, Oxford OX2 6DP
Oxford New York Toronto
Delhi Bombay Calcutta Madras Karachi
Petaling Jaya Singapore Hong Kong Tokyo
Nairobi Dar es Salaam Cape Town
Melbourne Auckland

and associated companies in
Beirut Berlin Ibadan Nicosia

Oxford is a trade mark of Oxford University Press

Published in the United States
by Oxford University Press, New York

British Library Cataloguing in Publication Data
Jones, D. S. (Douglas Samuel)
Assembly programming and the 8086 microprocessor.
1. Intel 8086 (Microprocessor)—
Programming 2. Assembler language
(Computer program language)
I. Title
005.2'65 QA76.8.I292
ISBN 0-19-853743-3
ISBN 0-19-853742-5 Pbk

Library of Congress Cataloging in Publication Data
Jones, D. S. (Douglas Samuel)
Assembly programming and the 8086 microprocessor.
Bibliography: p.
Includes index.
1. Intel 8086 (Microprocessor)—Programming.
2. Assembler language (Computer program language)
I. Title.
QA76.8.I292J66 1988 005.265 87-22031
ISBN 0-19-853743-3
ISBN 0-19-853742-5 (pbk.)

Set by Macmillan India Ltd., Bangalore 560 025
Printed in Great Britain by
at the University Printing House, Oxford
by David Stanford
Printer to the University

PREFACE

This book arose because I wanted a micro to do something which the high-level languages available did not have the facility to do. The only alternative was to write my own program in assembly language. It cannot be said that the books I consulted about assemblers were straightforward reading. Accordingly, I constructed a series of notes as I went along to help with my understanding. These notes, supplemented by some account of general principles, have been useful in other contexts. It is hoped, consequently, that what has been written will be found helpful to other amateurs, and maybe even a few professionals, concerned with programs in assembly language. While assembly language is specific to a particular processor, in this case the 8086, it is hoped that some of the descriptions of the environment which the programmer has to cope with will be generally useful.

My thanks are due to Dr. I. T. Adamson for a suggestion which improved Chapter 4. I am also grateful to Mrs. D. Ross for the efficient manner in which she converted my handwritten scrawl into an elegant typescript. My debt to my wife Ivy is immeasurable; she continues to display remarkable love and patience despite the burden that the preparation of the manuscript occasioned.

Dundee D.S.J.
October 1987

To those who have played a large part in my life

Ivy
My mother, Dot, Joyce
Helen, Philip
Katie, Kim, Corrie

CONTENTS

1. Generalities	1
2. Getting started	20
3. Cycling	43
4. Subroutines	58
5. Strings	76
6. Binary operations	90
7. Interrupts	109
8. Communication	121
9. System calls	138
REFERENCES	177
Appendix A: ASCII codes	179
Appendix B: 8086 mnemonics	183
Appendix C: Conditional jumps	192
Appendix D: Assembler directives	193
INDEX	197

1
GENERALITIES

Lots of jargon recurs in connection with microprocessors. Reading the literature is thus a chore for those unfamiliar with it. But some knowledge of the jargon is unavoidable if you are to fathom the meaning of the written word. So this chapter collects together some of the terminology that you may meet and gives a brief explanation of its significance. Such collections do not make for pleasant reading because the absorption of notions in rapid succession divorced from real applications is not something that most of us take to kindly. Nevertheless, even a fleeting acquaintance with some of the ideas is helpful. My advice is to read as far as you find tolerable and skim over the rest, stopping for any interesting morsel, so that you acquire a broad-brush picture of the content. This should be enough to enable you to begin on the programming in assembly language in the next chapter; a total grasp of everything in this chapter is not essential in the early stages. You can always come back to this chapter when an unaccustomed term turns up.

Architecture

1.1 Introduction

Microprocessors occur in numerous places. They are at the heart of personal computers, word processors, and small business computers. They also occur as controllers in such things as washing machines (to adjust the cycle), cars (to optimize engine running or achieve maximum braking without skidding in icy conditions), cash dispensers, automatic cameras, and many appliances where it is desirable to take account of several features while controlling in a reliable manner.

Generally speaking, a computer consists of four connected parts, each of which has a distinctive role to play. The four parts of a typical computer are displayed in Fig. 1.1. The key element is the *central processing unit* which has three functions to perform:

(a) to control the operation of the whole system by issuing appropriate signals to all parts of the machine;
(b) to receive instructions from any program stored in the memory and act upon them;
(c) to carry out all arithmetical and logical calculations, and to manipulate data.

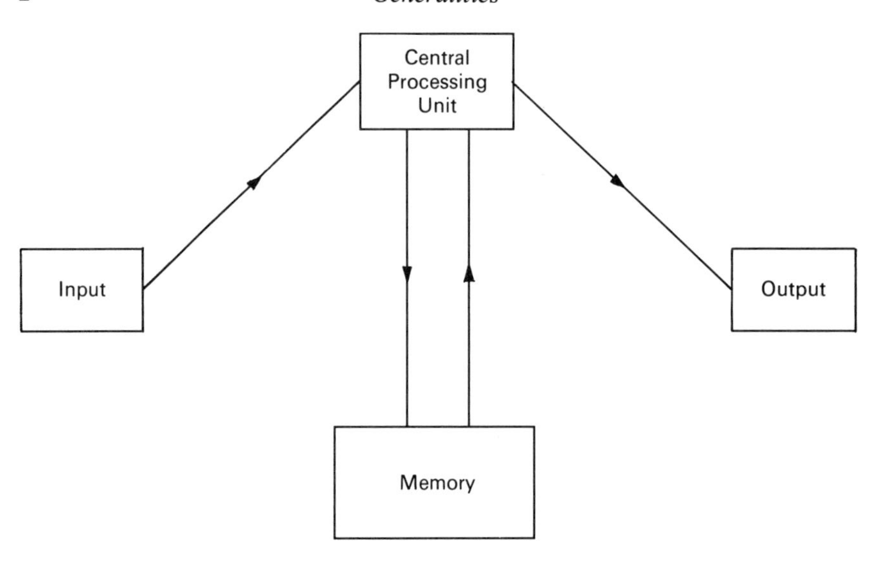

FIG. 1.1 Global architecture

Commonly, the term microprocessor refers only to the central programming unit, and we shall follow this practice. Thus, a microcomputer will be understood to have all four elements of Fig. 1.1, whereas a microprocessor will signify the central processor. To put it another way, a microprocessor is a device for controlling the memory and peripheral equipment, capable of following instructions from the memory and undertaking calculations.

Storage of information is the responsibility of the box marked *Memory*. Obviously, this information must be readily availabe to the central processor. Also, since the memory contains both data and instructions, we need to be sure that the central processor correctly identifies them when consulting the memory. To help with this each cell of the memory is given an *address*. When the central processor wants information from that cell it sends the address with a request for the content. The memory meets the request by copying the content of the addressed cell and sending the copy to the processor. In this procedure, the *content of the memory cell is unaltered*. If we want to change what is in a memory cell we must instruct the central processor to write data to the cell.

The *Input* is a port which passes electrical information in a suitable format to the central processor. It might be a keyboard, a disk drive, a modem gathering information from a telephone line, or a line carrying the current temperature. Similarly, the *Output* is a port to which the central processor sends information to be converted into a convenient format. It might be a display monitor, a disk drive, a printer, and so on.

1.2 Registers

There must be places where the input can put its data for subsequent transfer to the central processor and memory. A place is also required for storing the results of a calculation in preparation for further computation or output. The central processor needs to know what its next instruction is and where to find it in memory. The special places which serve these purposes and others are known as *registers.*

Basically, registers fulfil one of two functions, namely (i) the temporary storage of data and results as a calculation proceeds, and (ii) the control of the central processor in carrying out a particular task. Within these two categories registers can be designated for certain special functions though some processors allow registers to be used freely for any purpose, i.e. they are not restricted to holding data or a memory address or an instruction or whatever. For those processors which do distinguish between their registers one needs to know the constraints on usage and some of these are indicated in the following paragraphs. A careful examination of the registers and their facilities is always necessary when encountering a processor for the first time.

A *data register* is intended to handle data. Its size must therefore be adequate to cope with the largest package of data that the processor moves around.

An *address register* is provided primarily to keep track of an address. It therefore needs to be big enough to hold a typical address.

The aim of an *index register* is to hold an address that will simplify the processing of a table, though sometimes an address register may also be labelled as an index register.

A *status register* records the state of various parts of the processor. Each bit of the status register is called a *flag*. In fact, the register is often termed a *flags register*. A flag is said to be *set* or *raised* if the bit is 1 and to be *down* or cleared if the bit is 0. For instance, you might have a flag that was raised to 1 if a calculation overflowed capacity and was down at 0 otherwise. The relation is conveniently remembered because raising is higher than lowering as 1 is greater than 0.

An *instruction register* naturally deals with instructions. Usually, the current instruction is kept here while it is interpreted by the processor and executed.

Likewise, a *memory address register* will be used to give the address of the memory that is relevant to the current operation.

There are some registers to which the specific label register is not attached. One of these is the *program counter* or *instruction pointer*. It is responsible for seeing that the program is followed in the correct order. It does this by telling the processor where the *next* instruction to be executed will be found in memory. When the processor has finished with its current instruction it refers to the program counter which passes on the information that enables the processor to bring out the next instruction in the program from memory.

Another name of frequent occurrence is the *stack pointer* whose description we defer until the section on stacks (Section 1.4).

1.3 Instructions

A *program* formally consists of a list of instructions. Although there are computers where several processes can be carried out in parallel we are concerned only with instructions that are implemented one after the other, i.e. they are executed sequentially. For the moment we are concerned with the procedure for executing an instruction and not with whether it is directed at memory or a register.

The first stage in the procedure is the *fetch cycle* during which the instruction is fetched from memory. As already explained, the address in the program counter or instruction pointer is used to find the instruction in the memory. The instruction is then copied to the instruction register where its meaning is worked out so that it can be executed at the next step. The fetch cycle is completed by advancing the instruction pointer so that it points to the next instruction.

The second step is the *execute cycle* during which the action required by the instruction is achieved. It may be nothing more than copy a memory location into a register or it may involve some arithmetic but, whatever it is, the processor does not move to the next instruction and its fetch cycle until the current cycle is finished.

Thus instructions are first fetched and then executed. Naturally, the instructions have to be loaded into memory before this process can begin. In other words a program is first loaded and then it is run. While it is running the processor is working through the instructions in a steady cycle of fetch, execute, fetch, execute, ...

Not all fetch cycles occupy the same time interval nor do all execute cycles. An instruction which fills a larger portion of memory than another will take longer to fetch than the other. Similarly, the more complicated an instruction the longer it will take to execute. Precise timings depend on the structure of the processor and its set of instructions.

1.4 Stacks

The *stack* is a special way of organizing the memory so that only one location is available for operations at any instant while maintaining a record of certain previous transactions. In a stack the memory locations are imagined to be piled on top of one another (see Fig. 1.2). The stack pointer is set to the top of the stack. The rule is that you cannot touch any part of the stack except the top. Thus, in Fig. 1.2, you may either remove the 18 or add another number above the 18. These are the only operations available with a stack. Usually, the instructions are known as PUSH and POP (or sometimes PULL). For

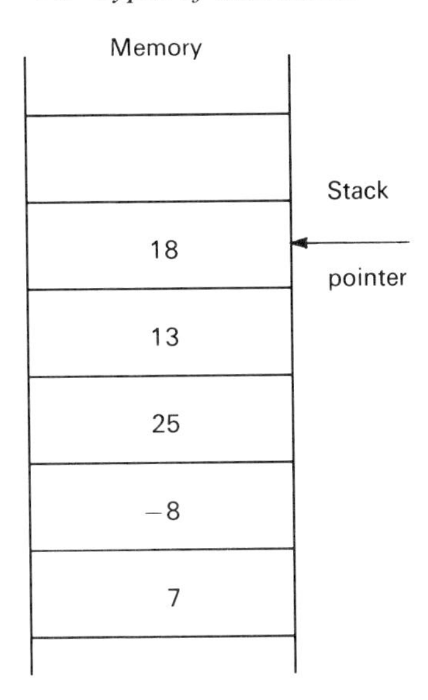

Fig. 1.2 Stack organization

example, if the location M contains 31 the instruction PUSH M copies the content of M to the position above 18 and moves the stack pointer up one (see Fig. 1.3(a)). On the other hand, if the instruction had been POP M, the content of the top cell (18) would have been copied to M and the stack pointer moved down one (Fig. 1.3(b)).

Stacks are easy to implement in memory because only a single pointer is needed. This identifies where the latest item is if we want it and also shows us where the next piece of data is to be put. There is the disadvantage that if you desire some earlier data you have to pop all the items above before you can get at it.

Instructions

1.5 Types of instruction

Numerous forms of instruction occur and there is no commonly agreed way of classifying them. Obviously, you expect every processor to have arithmetical, logical, and shift instructions, but beyond that it is more a matter of taste whether one prefers a general classification to one with finer detail. In deciding whether the set of instructions of one processor is more appropriate to your purpose than that of another one must beware of comparing the number of instructions without studying their impact. A larger number of instructions

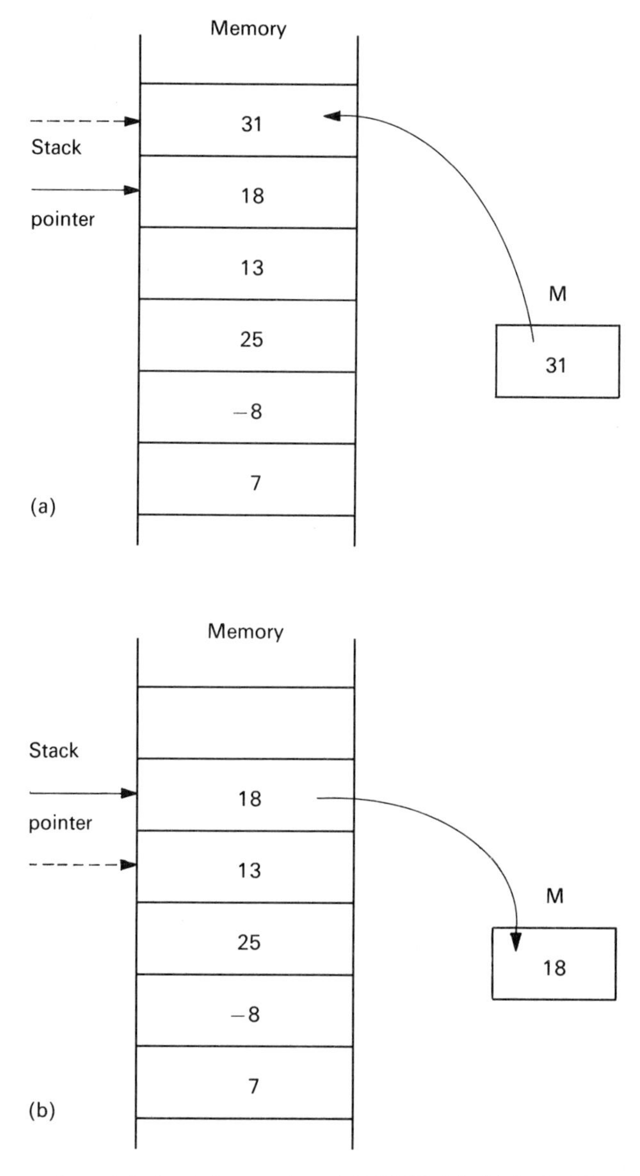

FIG. 1.3 Stack instructions (a) PUSH M, (b) POP M

may make one processor more flexible than another. Against that must be set the fact that the more powerful instructions are the smaller the number of them that have to be provided. Also, the time that elapses in undertaking certain operations may be a relevant factor.

1.5.1 Arithmetic Clearly, you will want instructions that allow you to add and subtract numbers. You would also like to be able to multiply and divide without having to construct them from repeated addition or subtraction. Further instructions may be supplied. Some of those which you can see might be useful are: changing the sign of a number, increasing a counter by one, decreasing a counter by one, adding with carry, and subtracting with borrow.

1.5.2 Logic Logical instructions draw conclusions from statements which may be either *true* or *false*. Since the operations are frequently applied to bits, the tables will be set out in terms of 1 and 0 which you can interpret as true and false respectively if you wish.

In NOT p every 1 is altered to 0 and every 0 is changed to 1 (Fig. 1.4). For this reason NOT p is sometimes known as the bitwise or *ones-complement* of p.

In p AND q two quantities p and q are connected. Each bit of p is matched against the corresponding bit of q. If both bits are 1 the result is 1, otherwise the result is zero (Fig. 1.5). An AND with a 1 reproduces the other bit whereas an AND with a 0 always makes the answer zero.

Bitwise matching also arises for p OR q. In this case the answer is 0 if both bits are zero but is otherwise 1 (Fig. 1.6). An OR with a 1 always gives 1 while an OR with a 0 reproduces the other bit.

For p XOR q there is again bitwise matching but this time the result is 0 if both bits are the same and 1 if they are different (Fig. 1.7). An XOR with a 1 changes the other bit but an XOR with a 0 leaves the other bit unaltered.

It will be noted that AND can be used to *mask* (i.e. make zero) bits of a

p	NOT p
1	0
0	1

Fig. 1.4 NOT p

p	q	p AND q
1	1	1
1	0	0
0	1	0
0	0	0

Fig. 1.5 p AND q

p	q	p OR q
1	1	1
1	0	1
0	1	1
0	0	0

Fig. 1.6 p OR q

p	q	p XOR q
1	1	0
1	0	1
0	1	1
0	0	0

Fig. 1.7 p XOR q

binary number, which permits easy checking of whether there is a 1 in a specified spot. Moreover, OR can be employed to merge two 8 bit numbers, one of which has 0 for its lowest four bits and the other 0 for its highest four bits, into a single 8 bit number without loss of information. Observe also that p XOR p always produces thc value zero.

Other logical operations can be considered but the four above are the ones usually present. If they can be combined in a single expression an order of procedure is necessary such as NOT, AND, OR, XOR; i.e. on a first scan the NOTs are evaluated, on the second the ANDs, and so on.

1.5.3 Shifts Three types of shift, in which bits are moved sideways in memory or a register, are generally provided. Sometimes the movement is limited to a shift of 1 position and sometimes the bits can be moved any number of positions. In some cases, the flags can be incorporated in a shift.

Rotate shifts (ROL, ROR) just push the bits round as if they were on a circle, the direction of movement in the memory being specified to the left or right (Fig. 1.8). Bits pushed out at one end are inserted at the other end.

The *logic shifts* (SHL, SHR) merely move all the bits in one direction and any bits which come out of the end are lost (Fig. 1.9). Zeros are inserted at the other end to fill up the vacant space.

Of the *arithmetic shifts* (SAL, SAR), SAL is the same as the logic left shift SHL (Fig. 1.10). This operation doubles the number with each shift of one

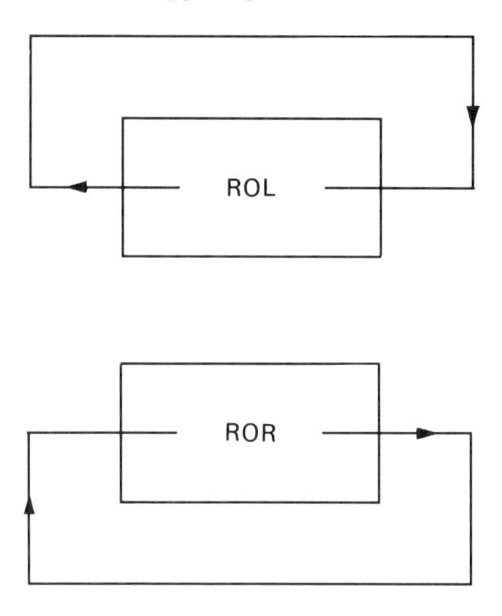

FIG. 1.8 Rotate shift

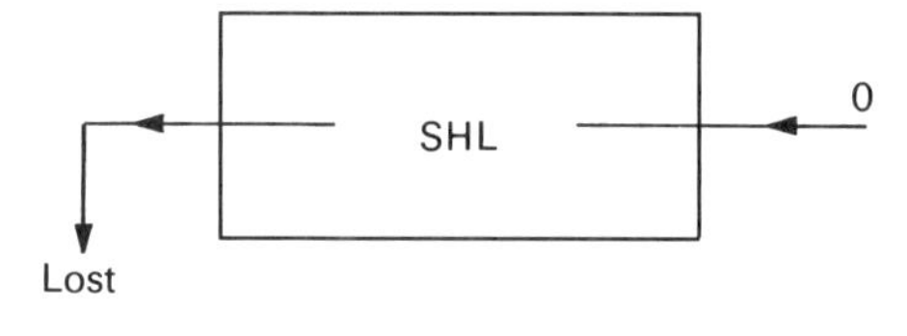

FIG. 1.9 Logic shift

position to the left (so long as the whole number is not moved out!) In the right shift a duplication is first made of the highest order bit and then copies of this are inserted as the number is pushed along. Thus, if the leftmost bit is a 0, zeros are inserted and, if it is a 1, ones are inserted. The reason for this operation is that the leftmost bit is customarily a *sign bit*, 0 indicating a positive number and 1 a negative one (see Section 2.10). The effect of filling the vacancies in the way described is to halve the number with each shift to the right.

1.5.4 Data transfer instructions Making transfers of data from one location to another is, of course, essential to the operation of a computer. An example

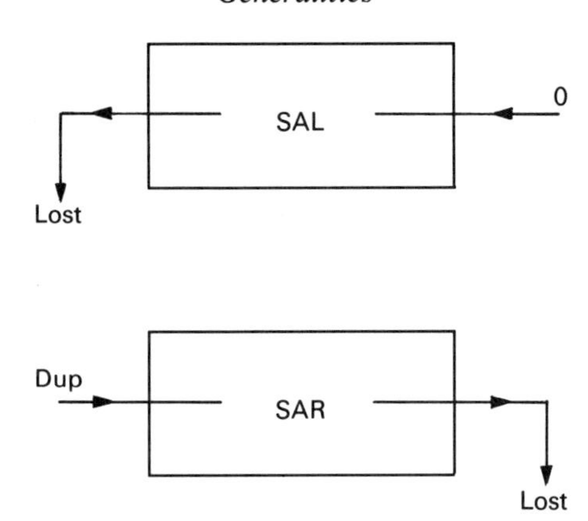

FIG. 1.10 Arithmetic shift

has already been met in the discussion of a stack. Obviously, there will have to be instructions which move, or rather copy, information from an input port to a register and from a register to an output port. Generally, there will need to be instructions which move the content of a given source to a specified destination. The instructions may be categorized as general purpose if source and destination can be either memory or register, but some instructions place restrictions on the source or destination or both.

Other instructions which might be encountered are the clearing of a memory location or register and the exchanging of the contents of two registers.

Special instructions for manipulating strings may be available. Often these are accompanied by a repeat instruction which permits repetition until some counter has fallen to zero.

1.5.5 Control transfer instructions All control transfer instructions cause the execution of the program to switch to a new place by modifying the program counter. As a consequence, when a control transfer instruction is met in a sequence of instructions, the processor is told not to take the next instruction in the sequence but instead to begin the instructions somewhere else.

Jumps (sometimes called *branches*) may be either *absolute* or *conditional.* For instance, 'Jump to the instruction in memory location X' permits no choice and the instruction pointer must point to X. In contrast, 'Jump on zero to the instruction in memory location X' checks first if the most recent previous result is exactly zero. If it is zero, (which will be recorded by the Zero flag being raised) the program will transfer to the instruction in X. If it is not zero (Zero flag down) the program will continue with its original instructions. In other words, a conditional jump instruction causes a jump if the condition is true but is ignored if the condition is not fulfilled.

Tests for conditions may be based on flags as already mentioned or on the comparison of two values. *Compare* instructions perform a subtraction of one value from another. They are therefore akin to arithmetic instructions. However, in a comparison the values are not altered and the result of the subtraction is not kept. All that happens is that a flag is raised or cleared depending upon the result of the subtraction.

Call instructions furnish a method of bringing in *subroutines.* A call instruction is actually an absolute jump to a subroutine. The new feature is that a *return* instruction at the end of the subroutine sends the processor back to the next instruction after the call in the original sequence.

Another type of instruction is the *interrupt* which is an asynchronous signal from outside the processor asking the processor to stop what it is doing and pay attention. This is analogous to the knock on the door by a visitor when you are busy reading. In the same way that you have to make special arrangements for your visitors, interrupts are handled by special programs such as *interrupt service routines.* When this routine is completed the processor is told to return to its original program at the point where it was interrupted.

1.5.6 Processor control instructions Processor control instructions are primarily to manipulate flag registers. They may also help with synchronizing the central processor unit with external hardware by, for example, telling the processor to cease operations until an interrupt or a certain signal is received. One common instruction in this category is NOP in which no operation is performed; the instruction merely occupies memory space and processor cycles but has no other task.

1.6 Types of address

Processors display great versatility in the way which they deal with addresses in instructions. Some of the principal modes that are deployed will now be described.

Consider the instruction

add the number 707 to the number 623.

In this case 707 and 623 are said to be *immediate addresses.* Strictly, they are not addresses at all but the actual values of the operands to be used in the execution and often the term *immediate value* is used. So, when the instruction is fetched, the actual values of the operands are fetched with it. No reference to memory is necessary. This mode is helpful in operations with fixed constants.

The next instruction to be considered is

add the content of memory location 707 to the
content of memory location 623.

Now 707 and 623 are called *direct addresses*, because they point directly to the memory locations that are to be used.

A more complicated usage arises as follows. Suppose that the memory location 707 contains the number 8 and that the memory location 623 contains the number 5. We might wish to issue the instruction

add the content of memory location 8 to the content of memory location 5.

This can be indicated by means of square brackets [ ] by adopting the convention that [M] means the content of memory location M. Then, our instruction can be expressed as

add the number in [707] to the number in [623].

A location such as [707] is said to be an *indirect address*, because 707 does not hold the operand but tells the instruction where the operand is.

The *effective address* is the place where the instruction actually ends up when it is executed. Thus the direct address 707 is also the effective address but, for the indirect address [707], the effective address is not 707; it is the address found in 707 i.e. 8.

There are similar modes for registers. For instance, a *direct register address* uses the operand contained in the specified register. Memory does not come into it at all. An instruction such as

copy Reg1

makes a copy of the content of the register designated Reg 1.

An *indirect register address* finds its effective address inside the quoted register. The instruction

copy [Reg1]

makes a copy, not of the content of the register Reg 1 but of the location whose address is in the register.

Index addressing forms a different mode because it creates the effective address by combining the information from two sources. The address given in the instruction is added to the address in the index register and the sum is taken as the effective address. It should be remarked that, despite being the effective address, the sum itself never materializes in any register. If I denotes indexing an instruction such as

copy 707,I

copies the content of the memory location whose address is 707 plus the number in the index register.

Associated with index addressing and with indirect register addressing may be the facility of *postincrementing* or *preincrementing*. In indirect register addressing with postincrement the effective address is discovered inside the register; however, after use, it is incremented by one and the new value placed in the register. With preincrement the address in the register is incremented by one and this increased value is employed as the effective address for the instruction.

Similarly, *postdecrement* and *predecrement* do a reduction of one after or before an instruction is implemented.

Base addressing is similar to index addressing except that a base register substitutes for an index register. The effective address is now the sum of the operand address in the instruction and the address in the base register. When the base register is a program pointer, base addressing is sometimes known as *relative addressing.* In this connection it should be remembered that the program pointer is always indicating the *next* instruction and this must be allowed for if the effective address is not to be in error.

Hardware

1.7 Memory

Up to the present, attention has been concentrated on how memory is used rather than its characteristics. Essentially, memory can be split into two portions; part is reserved for the manufacturer and the rest is at the disposal of the user.

Often the programs that are responsible for controlling the system are stored permanently. Usually, this is achieved by *Read-Only Memory* or *ROM* for short. ROM cannot be changed by the microprocessor so that the user cannot alter it either deliberately or by accident. It can, however, be read or copied like any other piece of memory. Naturally, the design has to be such that ROM retains its content in the absence of any power supply.

Versions of erasable ROM are available for some purposes. *EPROM* (Erasable Programmable ROM) can be programmed by the user and is then permanent until the user desires an alteration. The program can then be erased by various techniques and the EPROM reused. A similar device is *EAROM* (Electrically Alterable ROM) but the difficulties of changing this are such that it is often regarded as ROM after it has been programmed.

The memory that the user has for storing programs and data is *Random Access Memory* or *RAM*, for brevity. RAM can both be written into and read from. No parts of RAM are preferentially treated, so that all locations are equally accessible to the user. Perhaps the reader should be reminded that disk or cassette storage does not provide RAM because the reading head may have to be passed over other data before it reaches the desired location.

1.8 Input/Output

Terminals, printers, and the like have to be connected to the microprocessor eventually. In particular instances the connection may on occasion be made directly. In general, however, the designer will wish to keep the microprocessor from being tied to a particular piece of hardware and permit a reasonable variety of peripherals. It is common, therefore, to insert between the microprocessor and peripheral a block of software known as a *device driver*. The transition through the device driver takes account of any peculiarities of the peripheral which it controls. The extent to which peripherals can be interchanged depends upon the versatility of the device driver. If a new peripheral is substituted which is not compatible with the device driver then another device driver has to be furnished. In other words, the device drivers manipulate hardware while offering a consistent interface to the microprocessor.

One important aspect that affects design is whether information is to be transmitted in parallel or in series. In *parallel transmission* there are several parallel paths along which information can be transmitted simultaneously. For example, with eight parallel channels eight bits can be despatched at a time. Parallel transmission is therefore fast because of the amount of information which can be despatched per unit time. The price to be paid is in the provision of multiple parallel circuits. Consequently, for electrical pulses the method tends to be confined to short distances. If the use of a single optical fibre is feasible there are complications at the transmitting and receiving ends but the limitation to short distances is removed.

In *serial transmission* the bits follow one another in sequence. It is therefore bound to be slower than parallel transmission. Its advantage is that it works for long distances. Feeding input to a computer along a telephone line by serial transmission is a commonplace operation.

1.9 The bus

The transfer of information between the central processor unit and the memory is usually accomplished by parallel transmission. Clearly, the speed of operation depends crucially on the number of parallel paths available. This link between processor and memory is often known as a *bus*. Traffic in either direction is permitted on the paths in a bus, i.e. data can go from processor to memory or vice versa on the same path. Any link joining parts of the system with a similar structure is also called a bus. So a system may have just one bus or several depending on the design and manufacturer—for instance, a separate bus might be involved for printing.

Rules and priorities for traffic on a bus have to be established. Generally, this is the responsibility of the processor. However, it may agree to pass control to another device which wishes to send information along the bus. Typically, this will occur when a peripheral has input ready to go direct to memory.

Programming features

1.10 Numbers

Number may be represented in many different ways depending upon the base, or radix. A *binary number* consists entirely of ones and zeros. An *octal number* is composed of figures drawn from the digits 0,1,2,3,4,5,6,7. Of course, a decimal number employs 0 to 9. A *hexadecimal number* uses the decimal digits plus the six characters A(10), B(11), C(12), D(13), E(14), and F(15), where the number in brackets is the decimal equivalent of the letter.

The following conventions are adopted. The most significant figures will always be the leftmost, as in normal decimal usage. The base of a number will be indicated by a letter at the end (B for binary, H for hexadecimal) except that decimals will not be given a trailing identifier. Thus 10B is the binary representation of the decimal number 2 and should not be confused with decimal 10. In order that hexadecimal numbers such as ADD should not be mixed up with an instruction any hexadecimal which starts with a letter should be prefixed by a zero i.e. the number ADD should be written 0ADD.

It should perhaps be pointed out that there is no unanimity about the way of expressing a hexadecimal number. It is not uncommon to use a prefix instead of a trailing identifier e.g. $1A and &H1A are both forms for the hexadecimal 1A (which is decimal 26). However, we shall adhere to the notation of putting H at the end. The following decimal equivalents are worth noting:

Hex	*Decimal*
80H	128
0FFH	255
0FFFFH	65 535
100000H	1 048 576

A *byte* consists of 8 bits. It is therefore capable of representing 2^8 or 256 different numbers. In other terms, a byte can always be represented by two hexadecimal symbols and vice versa. It is customary to refer to 2^{10} bytes or 1024 bytes as a *kilobyte* and 1 048 576 bytes (2^{20} bytes) as a *megabyte*. To cover all numbers up to a megabyte requires 28 bits or 7 hexadecimal symbols. However, it is standard practice to stay with bytes because processors tend to work with multiples of a byte. Following the pattern you will guess that 64 kilobytes are, in fact, 65 536 bytes. Abbreviations for kilobyte and megabyte are kb and Mb respectively.

The group of bits which occurs in a normal transfer between processor and memory is usually referred to as a *word*. Thus, in a 16 bit machine a word contains 16 bits or 2 bytes, which means that a word can represent 65 536 numbers. The processor may, of course, have the flexibility of being able to move a single byte instead of never being able to handle less than a word. A collection of 16 bytes or 8 words is sometimes referred to as a *paragraph*. Larger groups such as 256 bytes or 512 bytes are occasionally known as *pages*, but there is no agreed standard and it is necessary to check the definition on each encounter.

One trap to be wary of is to think that when you press the key marked 2 on a keyboard the number 2 is sent to the processor. In fact, each key has a byte associated with it and when a key is pressed its corresponding byte is despatched. The machine then checks in a table to find out what this byte signifies. This has the advantage that by changing the code of the table you can adapt the keyboard to your own purposes though not all machines permit alteration of the decoding table. The standard table that is usually provided is based on the ASCII code (see Appendix A). From the list in the appendix you will see that the key marked 2 must send the byte with decimal value 50 for the machine to make a correct identification. Similarly, the byte with decimal value 97 will be interpreted as the letter a. Since keyboards rarely possess 256 keys some inputs need two or more keys to be pressed in combination so that the decoding table recognizes the result correctly. You will observe that lower case letters differ from upper case counterparts only by a single bit, which can be made 1 or 0 by an appropriate operation. If, for some reason, you actually wanted to send the byte for decimal 2 (which is a control code) you would have to press the control key and B or ^B as it is normally written, though not all manufacturers adopt this convention.

While you can expect to find the first 127 sybmbols of the decoding table readily available from the keyboard the same may not be true of the remaining characters. Some manufacturers use the extra slots entirely for continental characters like ü, ô, or the Greek alphabet while others may offer access only after special instructions have been issued since many of the symbols are related to modes for graphics.

Occasionally you will meet the terminology that some information has been formatted as an *ASCIZ string*. This means that the string consists of ASCII characters terminated by a byte of 00H.

1.11 Languages

Programs consist of sets of instructions. The instructions have to be in an agreed language. The *level* of a language is determined by the distance it is from the machine's own language which is regarded as the *lowest level*. The machine's own instructions are written in bits and, although a user could be asked to convert all instructions into equivalent binary representations, such an arrangement would scarcely be regarded as friendly. So *high-level languages* which are more akin to English have been developed. Typical examples are BASIC, FORTRAN, PASCAL, ALGOL, and ADA. With these the user has no knowledge of what is going on in the processor and, moreover, does not care so long as it performs as desired. Naturally, a special program, called a *compiler*, has to be provided to turn a program in a high-level language into instructions which the machine can recognize. Again the user is not concerned with the details so long as the performance is up to scratch.

At a level between the machine and high-level languages is *assembly language*. In assembly language each instruction corresponds generally to one operation of the computer. The instructions are written in mnemonics so that one can name addresses rather than having to specify precise locations at every stage. In order that the computer can understand this language there has to be a special program called an *assembler* which converts the mnemonics and addresses into a form which the processor can obey. The assembler therefore has an analogous role to the compiler for a high-level language. But whereas a manufacturer always includes a compiler with every high-level language supplied, not all makers provide an assembler automatically and it may have to be obtained as an extra item.

Programs in assembly language contain instructions of two types. One type is an instruction which the machine has to follow. The other type is an *assembler directive* (sometimes termed a *pseudo-op*). A directive to the assembler tells it how the program is to be translated into machine code, but it is never passed on to the processor because the assembler completes its duties before the program is loaded and executed. For example, an assembler directive might be to allocate some memory for the result of a calculation, leaving it to the assembler to choose where. After assembly is over a decision will have been made on the location and the machine receives an instruction which tells it exactly where the result is to be placed.

On account of the two types of instruction not all assembly programs for the same purpose on the same central processor are the same. The machine instructions must be the same and be in the mnemonics given by the designer of the central processor. The assembler directives, in contrast, must fit the rules of the designer of the assembler and, since there is no standard assembler directive, these can vary from supplier to supplier.

For this reason the assembly programs which occur later may not always contain the directive you are employing, but it is hoped that the minimal use which is made of them and the explanation which is offered will enable a transition which is relatively smooth (with the help of Appendix D).

A natural question is to ask why one should bother with assembly language. There are several answers. You may wish the computer to undertake action which is outside the province of your high-level language; in that case you have to write your own machine program. In some circumstances your high-level program is consuming so much memory and time on the computer that the saving achieved by converting to assembly language justifies the effort of reprogramming. Again, you may wish to deploy the processor in a situation where high-level languages are not available e.g. where space and memory are at a premium. Finally, your interest may be that of becoming more familar with the operation of processors and the principles of programming.

Having said all that it is best to adhere to the criterion of choosing the highest level of language appropriate to your task. The program will usually be easier to write, easier to understand, less prone to errors, and involve less time on debugging. Also the program may be transportable to other computers. Assembly language entails close attention to details, and debugging, even with the special debuggers supplied by some manufacturers, is generally slow and tedious—a small error may lead to unexpected and seemingly inexplicable effects; sometimes the program may seem to have disappeared into limbo!

The preparation of a program first entails writing it in assembly language, sometimes called producing the *source code*. The program is then entered into the computer by means of a wordprocessor or an editor such as EDLIN or ED. This is in contrast to what happens with a high-level language like BASIC which includes an editor that is available for entering programs as soon as BASIC is loaded. Assemblers, generally, do not contain editors.

Once the source code has been entered the assembler is invoked. Its input is the source code and its output is the machine code, also known as the *object code*, which flows from the source program. It may also indicate any errors or illegal instructions which it has detected during the translation. When any mistakes have been corrected (which means a return to the editor followed by re-assembly) the object code is fed into a *linker*. The linker adds information required by the operating system like the size of the program, where it is to be located in memory and so forth. The linker also enables the combination of several object programs into a single one as well as being responsible for attaching any object programs which the operating system insists should be associated with any new material.

The output of the linker is a program in machine language which can be executed. It is this output which is preserved on disk and run by the operating system at the appropriate time.

This mechanism is quite different from the way in which a BASIC program is translated into machine language. The duty of an assembler is to see that the source code is free of syntax errors and to generate a complete executable object program before any run is undertaken. With BASIC no such program in object code is created. Instead BASIC, *as an interpreted language*, takes one instruction at a time from the source code, converts it to machine code and then executes the machine code before passing to the next instruction in the source. If it fails to carry out the requirements of the machine code for some reason it stops interpreting, issues an error message, and hands back control to you.

Compiled languages (e.g. PASCAL, COBOL) are a sort of half-way house. The compiler does not have the task of running the program. It translates each source instruction into a number of instructions in machine code ready for subsequent linking and running.

1.12 Good programming

Good programming resembles an elephant in that it is difficult to describe but much easier to recognise whether it is present or not. It cannot be learned from generalities or just from reading a book; it needs careful study, constant practice and imitation of methods which you observe to be effective. These is no substitute for personal experience in writing programs. Often a balance has to be sought between meticulous attention to detail and keeping the overall aims in mind. Obviously, the final program must give the right answers, run reasonably economically, and be capable of future modification.

Basically, when facing a complex problem, one breaks it down into several soluble sub-problems. Each sub-problem may then be split into several soluble sub-sub-problems. The process of subdividing, sometimes known as *refinement*, continues until a level is reached where the solution can be written down directly. All these steps should be documented systematically otherwise the separate portions may not fit together at the end. Avoid especially terse descriptions which you may not understand when you return to them.

Some criteria to think about adopting for making subdivisions are:

(i) Try not to have more than 4 or 5 subdivisions to a problem or sub-problem otherwise you will lose track of them.
(ii) Try to make the subdivisions as independent of one another as possible otherwise an alteration to one can have widespread effects on others.
(iii) Defer getting down to detail as late as feasible so as to avoid tying up the design prematurely and causing the structure to become rigid. Do not be afraid to re-think if things seem to be getting into a mess.
(iv) If you have found a good method for tackling a problem and the problem turns up again repeat the method unless there are strong arguments to the contrary (when you should leave yourself a note to remind yourself of why you departed from the previous pattern).

Exercises

1. A stack occupies memory. What is likely to happen if you make a large number of (a) PUSHs without any POP, (b) POPs without any PUSH?
2. As you drive towards a traffic light it changes to yellow and a pedestrian starts to cross the road in front of you. Regarding these as interrupts, decide on your priorities. Would you change your priorities if (a) a car was crossing the intersection in addition, (b) the pedestrian took one step and then stopped?
3. Do you have any preference for any of the shifts in Section 1.5(c)?
4. Would it be an advantage if every instruction had the length of a word?
5. Do you think it is helpful to have the NOP instruction? If so, why?
6. If the content of the register Reg 1 is 861 what is the effective address in (a) copy Reg 1, (b) copy [Reg 1] and what is actually copied? If preincrement is applied in (b) what is the effect?
7. The bus of a microprocessor has 10 paths. How many memory locations can be addressed? If each location contains (a) a byte, (b) a word what size of memory can be addressed? What would be your answers when the number of paths on the bus is (i) 8, (ii) 20?

2
GETTING STARTED

The Intel 8086 microprocessor

2.1 Facilities

The microprocessor on which we are going to illustrate assembly programming is the Intel 8086 (known as the *iAPX 86/10* in some quarters). Another popular processor which we might have chosen to work with is the Motorola 68000. Unfortunately, space restrictions prevent a full treatment of both and their structures are quite different. The more recent 80286 and 80386 are also omitted from consideration. While these can be used by programs in a mode compatible with the 8086 you should beware a direct transfer until you have confirmed that there will be no undesired phenomena. The fact that processors can be so different means that you cannot proceed to assembly programming until you have some familiarity with the facilities of the machine. You need to know the quantity of memory, its addressing, the set of instructions, and how these things are organized. Speed of operation and the arrangements for input and output may also be relevant. We will now try to describe the salient features of the 8086 while attempting to keep the amount of information down to digestible size.

The 8086 is a silicon chip based on metal-oxide-semiconductor technology. Such chips are highly reliable though not necessarily having the speed of other types. Connection is via 40 pins, some of which are for two purposes and can switch between applications. Because the pins serve two functions additional devices have to be provided outside the chip to handle the bus in any necessary switching. However, the number of outside chips depends upon whether the processor is in minimum or maximum mode (which is determined by whether a certain pin is at earth or +5 volts). In the minimum mode the processor retains control of the bus and external support chips are mainly devoted to memory and interfaces. In contrast, external chips play an intensive part in bus control in the maximum mode. Outside means of coping with the sharing of functions of the address and data buses are inevitable in any case.

The address bus has 20 paths and so can address 1 048 576 memory locations. With each location containing a byte the processor can cope with a memory size of 1Mb. (see Section 1.10). Your microprocessor may not be supplied with that quantity of memory. However, as you will see later, difficulties could arise if the physical memory was less than 256 kb. Bus management is carried out by a *Bus Interface Unit* which is responsible for fetching instructions, controlling input and output, and communicating with memory. An *Execution Unit* has the duty of executing instructions.

The Bus Interface Unit divides the memory into segments (which can overlap) and four segments can be involved in the execution of a program. On account of the overlapping in the fetching of instructions it is not easy to assess the speed of the processor. The best technique is to test it with benchmarks. An upper and probably unrealistic limit can be obtained by adding together the times of separate instructions. These timings are determined by the processor clock and how many cycles of the clock are needed. For example, a copy of a word from memory might take four cycles, while multiplication of words may require 100 or more cycles. The clock of the 8086 is customarily set to run at between 5 and 10 MHz. It is important to realize that it is the combination of clock setting and operation which fixes the overall speed. A processor with a fast clock which takes many cycles for every operation may have an overall speed which is lower than one with a slower clock which completes sophisticated operations in a few cycles. We emphasize again that proper benchmarking gives an acid test.

Another point to notice in the assessment of speed is the location of a word in memory. Although there is no restriction on where a word may be placed execution time is somewhat longer if the word does not start at an even address.

Input and output have separate controllers and interfaces i.e. they are regarded as distinct from memory. Therefore, special instructions are necessary for input and output.

The 8086 can handle data in bits, words, or strings, the instruction telling the processor which type is involved. There are special instructions for processing character strings. The 8086 can manipulate *doublewords* (32 bits) in a very limited way and can also deal with *binary-coded decimal*, but if you want floating point arithmetic then you will have to prepare a program.

2.2 Memory segments

Memory addresses are always regarded as increasing to the left in the same way as we put the most significant figures of a number on the left (see Fig. 2.1). As already mentioned the Bus Interface Unit splits the memory into segments. Each segment is 64 kb long and must start at an address which is an integer multiple of 16 bytes. In Fig. 2.1, A, B, C, and D are four possible segments. Remark that, since 4 segments are in operation at any one time, they can be fully utilized only if the physical memory has a minimum size of 256 kb.

The position of a memory location such as that marked by a cross in Fig. 2.1 is determined by first giving the starting point of a segment containing it (say A) and then listing its distance (called the *offset*) from A. Three different ways of finding the position, using segments A, B, and C are shown in the diagram. The

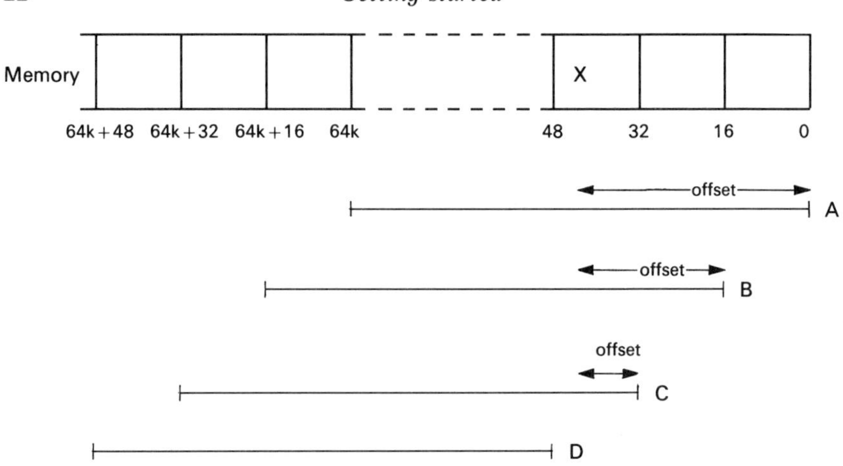

FIG. 2.1 Memory segments. Figures indicate address in bytes

cross cannot be referenced by the segment D because it lies outside the segment.

To keep track of a memory location we need to store to distinct numbers, the start of a segment and the offset. By virtue of always beginning at an integer multiple of 16 bytes the four lower bits of the address of the segment start are always zero. In hexadecimal the address will be of the form 12AB0H with the least significant character always zero. Hence, there will be no loss in stripping off the 0 for storage since it can be restored when we bring the address out of store. So the start of the segment is stored as 12AB and thereby occupies a word. As for the offset it can never exceed 64 k, the length of a segment, and so it too can be stored in a word.

If the offset is 213AH when the start of the segment is 12AB0H the memory location concerned is obtained by adding together these two numbers i.e. it is at 14BEAH. Observe that the final address obtained is a 20 bit number which, consequently, matches the size of the bus exactly. Thus, although the address is stored in 16 bit words which are smaller than the bus, the restoration of the zero at the end when the segment start is pulled out ensures that the final address is compatible with the bus.

A special notation is employed to indicate the above procedure. The address is recorded in the form 12AB:213A showing the numbers actually stored. When they are combined to make the address you must remember to put the final 0 on the first number. Symbolically, the notation is

segment start without last zero:offset

the numbers being hexadecimal. More briefly, we could write it as SS:OF. In general WXYZ:RSTU is interpreted as the sum of WXYZ0 and RSTU.

Arithmetically, the insertion of 0 at the end of the word stored for the segment corresponds to multiplying the word by the decimal number 16 because the zero is in hexadecimal. Similarly, when the hexadecimal 0 is stripped off the segment address for storage, the operation is division by decimal 16. Thus, in arithmetical terms,

$$\begin{aligned} 1234{:}5678 &= 1234H*16 + 5678H \\ &= 1234H*10H + 5678H \\ &= 12340H + 5678H = 179B8H \end{aligned}$$

Before leaving the topic of memory there is another important observation. The 8086 stores words following the same convention that we have for writing down numbers i.e. the most significant bits are on the left. Since addresses increase to the left this means that the high order byte has a higher address than the low order byte. If one is working with words this has no particular significance but, if one is pulling out a succession of bytes by steadily incrementing the address, it does mean that you will extract the low order byte of a word before the high order. It is something to watch out for if switching operations from words to bytes.

2.3 Registers

The 8086 processor has 14 registers each of 16 bits. They are displayed diagrammatically in Fig. 2.2 with the most significant bits on the left in accordance with our convention. Each register except that for flags is identified by a pair of letters as shown in the figure. The principal purpose of each register is also indicated but, in some cases, other usage is permitted.

You will notice that the top four registers, which are essentially for the manipulation of data, have a dashed line splitting them into high and low order bytes. Each of the bytes has its own pair or identifying letters. This is Intel's cunning device for allowing you to manoeuvre with words or bytes and tell the assembler which merely by the register designation. Thus, when a transfer is made to AX, a word of 16 bits is moved. A transfer of AH moves a byte of 8 bits and has no effect on AL. Accordingly, the top four registers can be thought of as four independent word registers or eight independent byte registers. They do not all have to be used in the same way; one can be for words and another for bytes.

In spite of the names of the four data registers all of them are available for arithmetical and logical operations with some exceptions. Multiplication and division of words can be done only in AX and DX; for bytes the corresponding registers are AH and AL. Counting of operations occurs in CX, with variable shift counts in CL. The port number for input or output will be in DX. Certain composite operations are numbered and this number is usually placed in AH. It is wise to check if an unfamiliar instruction is destined for a specific register.

The Stack Pointer is for the manipulation of the stack. It is worth noting

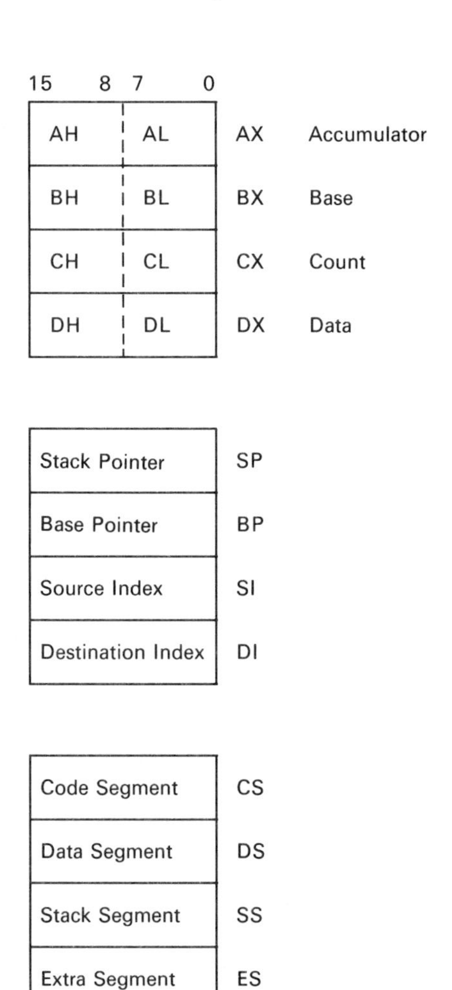

FIG. 2.2 Registers

that the bottom of the stack has a higher address than the top i.e. in Fig. 2.1 the bottom of the stack is on the left and the top on the right. When the operation PUSH adds an element to the stack the pointer moves to the right. The stack expands to the right and contracts to the left.

The Base Pointer is essentially at the disposal of the user either as a pointer or for data. Likewise, the index registers SI and DI can hold data, though in string operations they have a special role (in conjunction with DS and ES).

The *segment registers* are the storage places of the starting points of memory segments described in the preceding section. The *Code Segment* of memory contains the program and its instructions; it is referenced by CS. The *Data Segment*, whose start is given in DS, contains the primary data structures and variables of the program. The stack lies in the *Stack Segment* and the location of the top of the stack is at SS:SP (i.e. put zero at the end of SS in hexadecimal and add on SP) because the SP register for the stack pointer holds the offset of the pointer. Typically the content of the *Extra Segment* will be data.

During the running of a program the actual address in memory is determined by the processor via segment and offset. The instructions in the program generally refer only to the offset. For example, if there is an instruction in the program to operate on the data in location LAMBDA the machine, when running, finds the data actually at DS:LAMBDA. Therefore, normally, the programmer does not need to worry about the values of the segment registers and can relinquish responsibility to the operating system. There are methods of managing the segments with assembler directives so that you can secure more than 64 kb for code or data but they are beyond the scope of this book. In the programs we shall write we shall usually ignore the presence of the segment registers. When you allow the operating system to select the values of the segment registers, the programs are said to be *relocatable*, because the program does not have to go in the same place every time it is loaded.

The instruction pointer tells us where the next instruction is or, rather, the offset of that instruction. Since instructions are in the domain of the code segment the actual address of the next instruction is CS:IP. However, programmers are rarely concerned directly with the instruction pointer but rely on its correct automatic adjustment in jumps, calls, etc.

The structure of the flags register is quite different from the other registers. Firstly, seven of the bits are reserved to the processor; they do not exist as far as

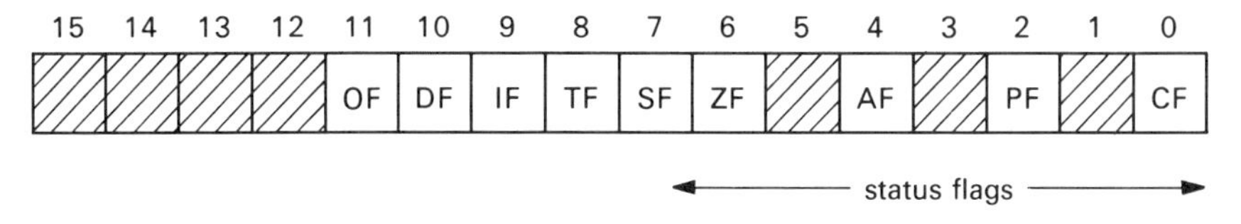

FIG. 2.3 Flags register

the programmer is aware and are shown shaded in Fig. 2.3. The remaining nine bits reflect the state of the processor. The programmer cannot gain access to the single-bit registers but can test them to discover the effect that the execution of an instruction has had because six of the flag bits (in the bottom 6 positions) can be affected by arithmetic operations. In other words, these bits supply *condition* or *status codes* which report what has happened in the most recent computation.

For example, if the result of the operation was zero the *Zero Flag* (ZF of Fig. 2.3) is set to 1 but, if the answer is non-zero, it is put to 0.

The *Overflow Flag* (OF) is raised to 1 if the size of the result exceeds the capacity of the destination.

The *Sign Flag* (SF) is set to 1 if the result is negative but is otherwise 0.

The *Carry Flag* (CF) is set to 1 if there is a carry from addition or a borrow from subtraction in the highest order bit, otherwise it is 0. What this means will be discussed in more detail later.

The *Auxiliary Carry Flag* (AF) is for binary-coded decimal (see Section 6.5) while the *Parity Flag* (PF) is set to 1 if the low-order eight bits of the result contain an even number of ones.

The *Trap Flag* (TF), *Direction Flag* (DF), and *Interrupt Enable Flag* (IF) are seldom of much interest in elementary programming and will not be described at this point.

2.4 The format of an instruction

In order to be able to follow examples we must first understand the structure of a processor instruction. Let us consider the following:

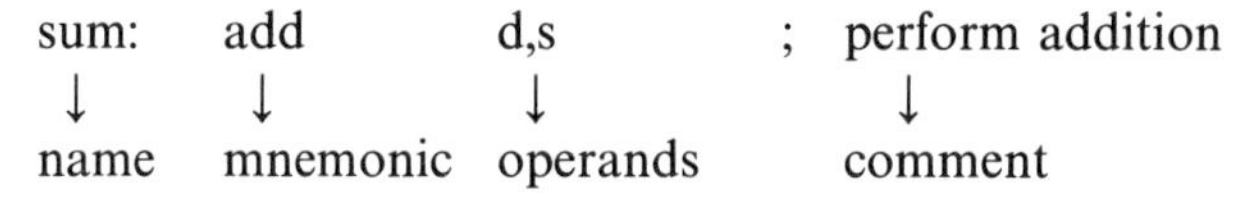

where the second line is a description of the line above.

Perhaps we should start with something that seems so obvious that it might be forgotten. Every assembly language statement should terminate with a carriage return and line-feed. The assembler then knows that the end of the line has been reached. Normally, you will insert the termination signal automatically via a text processor when writing your program into the machine. If you want multiple instructions on the same line see Section 7.5.

The first item, *sum*, which may occur in an instruction is a *name* or *label* which, preferably, should give an indication of the process you are about to start. It is *not* necessary for every instruction to have a label but any that the program is going to jump to must be labelled. Generally, labels are omitted whenever possible. You can choose a label to suit your convenience so long as it starts with a letter of the alphabet (note that most assemblers do not distinguish between upper and lower case letters except in specially designated strings). Subsequent characters can be alphabetical, numerical, or the *underscore* _. The underscore can be useful in improving readability e.g. my_label—observe that *you cannot replace the underscore by a space*. You should not, of course, pick a label which is the same as an assembler directive, a machine instruction mnemonic, the name of a register, or any other reserved word.

Usually there is a restriction on the maximum size of a label. Since you will wish to keep labels short this is not likely to be much of a constraint but it is wise to check whether you are limited to 10, 30, or 80 characters. The final colon, following immediately after the last character of the label, is essential for a machine instruction. For an assembler directive the colon is omitted but the label must be terminated by a space.

The *mnemonic* is shorthand for the operation you want carried out and therefore sometimes termed the *opcode*. In the example it is an instruction for the processor to perform, but it could be a directive for the assembler. Machine instructions must be standard 8086 mnemonics (or such modifications of them as have been introduced by the assembler) and a complete list is set out in Appendix B, though we shall bring them in only gradually. Assembler directives must accord with the code laid down by the assembler.

The *operands* are, in effect, part of the instruction indicating the objects that are involved in the operation. There may be two operands, as displayed in the example, or one or none. Any operand must be separated from the mnemonic by at least one space; if there are two operands they are separated from each other by a comma. Of two operands, one is designated the destination (d) and the other the source (s); the d always comes before the s. Thus, in the example, the result of the addition will appear in d. There may be restrictions on the addresses permitted for the source and destination operands as can be seen from Appendix B.

A semi-colon starts the *comment* and should be separated from the operand(s) by at least one space. The comment ends with a carriage return and line feed. Comments can be omitted but this is bad practice. A well-documented program should use comments liberally to help with debugging, to facilitate modifications in the future and generally assist the reader to understand what is going on in the program. A whole line can be devoted to a comment by starting the line with a semi-colon. Notice that a comment is always the last statement on a line because once the comment begins the assembler assumes that it extends to the end of the physical line.

Simple arithmetic

2.5 Addition

We will commence by considering the addition of a decimal number, say 17, to the 8 bit byte in AL which is actually the lower byte of the word in AX. We do not ask how the number got into the register but presume it was put there by another part of the program. Nor do we enquire what the program is going to do with the result of the addition. In other words we are concerned with a *program fragment* i.e. a part of a larger program, but not a complete program. Normally, the instructions of a program fragment will have some dots placed above and below to remind you that a full program will have further instructions.

In displaying programs and program fragments the convention that will be adopted is that those in assembly language will be in lower case and those in a high level language will be in capitals. In the descriptive text the names of registers, memory locations, instructions, labels of instructions, etc. will be in upper case in order to keep confusion to a minimum. For the same reason, similar references in comments in displayed programs will be in upper case while the main body of the comment will be in lower case.

Although AL is part of AX it is distinct in the sense that a byte operation on it has no effect on the remainder of AX, i.e. AH. In fact, the instruction infers from the presence of the register AL that the operands are in bytes and that the addition is in terms of bytes. Since 17 is to be added to the contents of AL, the destination operand is AL and the source operand is 17. Therefore the fragment of program looks like

```
. . .
add     al,17     ; add immediate value 17 to
                  ; the byte in register AL
. . .
```

Notice firstly that we have not bothered to attach a label to this instruction. Secondly, there is an explanatory comment and that this has been allowed to run over to a second line by making the semi-colon the first symbol of the extra line.

Since 17 is a fixed number and not a memory location its value goes along with the instruction without any reference to memory. That is why it is called an *immediate value*. Correspondingly, the addressing mode of this operand is said to be immediate. On the other hand, the destination operand is obtained directly from the register AL and so the addressing mode is direct register.

It will be observed that the number originally in AL is obliterated by the operation and cannot be recovered unless there is a copy somewhere else. Also, it may happen that the number in AL is so large that, when 17 is added, the result requires more than 8 bits to represent it. You might think that either the machine refuses to carry out the addition or that the result is allowed to overflow into the higher byte AH. You would be wrong on both counts. Instead, after the addition has been performed AL contains the lowest byte of the result, AH is unaltered, and a flag is set to warn of error in the answer. More will be said about this later.

Now consider the problem of adding the byte in memory location 1234H to the byte in 4321H and placing the result in 4321H. An attempt to achieve this by: add 4321, 1234, would fall foul of several errors. Firstly, the omission of the H would entail the interpretation that the numbers were decimal instead of hexadecimal. Secondly, even if H were inserted, the numbers would be treated as immediate values and not as the addresses of memory locations. Thirdly, the ADD instruction does not permit the addition of one value in memory to another number in memory (or of two immediate values).

What we must do is to copy one of the numbers into a register, carry out the addition there and then transfer the result to memory location 4321H. The instruction for copy is MOV, indicating that a value is being moved from one place to another. So we might plan to open with

```
mov   al,1234H
```

However, we have now fallen into another trap. The instruction will regard 1234H as an immediate value and believe that it is being requested to put 1234H into AL (which it cannot do, of course). So the assembler needs additional information which directs it to the byte in 1234H and this is provided by adding BYTE PTR where ptr is an abbreviation for *pointer*. (If we wanted a word rather than a byte we would substitute WORD PTR.)

Let us try again:

```
...
mov    al,byte ptr 1234H  ; copy byte from location
                          ; 1234H and place it in AL
add    al,byte ptr 4321H  ; add byte from location
                          ; 4321H to the byte in AL
mov    4321H,al           ; copy the result to
...                       ; location 4321H
```

Observe that we have had to use the pointer with ADD for the same reason as with MOV. Since MOV copies, there is still a copy of the result in AL as well as in 4321H. The number originally in 4321H is, however, irrecoverable.

While the above fragment does attain the desired object it reminds us also that if we use numerical addresses throughout our program we shall face a tedious task if one of them has to be changed and simple transposition errors may be tiresome to detect. It is therefore better to follow the example of high level languages and use names were feasible, though the significance of the name in numerical terms has to be described somewhere. Suppose we agree that 1234H should be named ONE and 4321H, TWO.

An assembler directive EQU incorporates these names in the program. Remember that an assembler directive is not translated into machine code; a directive simply tells the assembler to perform certain functions. In the case of EQU the assembler is told that something is equivalent to something e.g.

```
counter   equ   five
```

There are three rules to comply with for EQU. There is no colon on the name being defined on the left. The name on the left cannot be redefined later by a subsequent EQU or other directive. The label on the right must have been defined already in an earlier statement.

Let us repeat the previous addition with names but change it to words instead of bytes. Then we shall have*

```
        ...
one     equ     1234H               ; give the memory location
                                    ; 1234H the name ONE
two     equ     4321H               ; give the memory location
                                    ; 4321H the name TWO
        ...
        mov     ax,word ptr one     ; copy word from location
                                    ; ONE and place it in AX
        add     ax,word ptr two     ; add word from location
                                    ; two to the word in AX
        mov     two,ax              ; copy the result to
        ...                         ; location TWO
```

After assembly this program looks exactly the same as the preceding one because the assembler, every time it meets ONE, substitutes the address 1234H, i.e. the assembler has been assigned the job of putting in the actual numbers. If we wish to alter the memory location, only one instruction has to be changed and the assembler does the rest. Since the assembler substitutes the address number for the name the pointers must be retained, otherwise we run into the problem with immediate values.

Because the EQU directive merely tells the assembler to replace one quantity by another no room has to be found for it in the object code. Nevertheless, it is capable of quite complicated substitutions e.g.

```
count     equ 5
buffer    equ 80
top       equ count+buffer
mov       al,count+1
mov       bx,buffer*2
mov       al,count+buffer
mov       al,top
```

Notice that each of the last two instructions would generate exactly the same object code. For some additional information see Appendix D.

We have already remarked that we lose the original content of 4321H with the above program. Suppose we wish to keep it and place the result of the addition elsewhere. Where it goes does not concern us, perhaps, so long as we

* All mnemonics are copyright of Intel Corporation 1986

can lay our hands on it. This is another case for a label leaving it to the assembler to settle on the precise address. Another assembler directive is called for since EQU requires us to fix the actual address. Let us look at the program first:

```
        ...
result  rs     2                   ; reserve 2 bytes (1 word)
                                   ; in memory (without initial
                                   ; value) for result
one     equ    1234H               ; give location 1234H
                                   ; the name ONE
two     equ    4321H               ; give location 4321H
        ...                        ; the name TWO
        mov    ax,word ptr one     ; copy from ONE to AX
        add    ax,word ptr two     ; add word in TWO to AX
        mov    result,ax           ; copy the sum to RESULT
        ...
```

The assembler directive RS asks for the reservation of space in the memory but the assembler decides where. The amount of space in bytes is specified by the number following RS but the assembler is not obliged initially to place any particular values in the reserved location. During assembly the label RESULT will be changed to the actual address that the assembler has allocated.

Unfortunately, there is no agreement between assemblers on the directive for reserving space. Another popular one is

```
result  dw 1 ?
```

Here the dw 1 denotes the reservation of 1 word of memory while the question mark implies that no initial values are embodied. For further information on this topic see Appendix D.

2.6 Subtraction

Subtraction is so similar to addition that only one example will be given. Take the byte in the register DL from the byte in memory location ANSWER and keep the result in ANSWER.

```
         ...
answer   rs     1            ; reserve 1 byte (without
                             ; initial value) for ANSWER
         ...
         sub    answer,dl    ; subtract value in DL
                             ; from ANSWER and leave
                             ; in ANSWER
         ...
```

2.7 Multiplication

The instructions for multiplication have two restrictions. One of the numbers must be in AL if it is byte or AX if it is a word. The other number, which is displayed as an operand, is not allowed to be an immediate value. So , if we want to multiply the byte in location 1234H by 89, we must either put 89 in AL or, if several multiplications by 89 are desired, place it in a register or memory (though operations with registers tend to be somewhat faster). Let us assume that we do not want to place 89 in AL and that the product (which can be a word) is to go in location 4321H. Then a possible program fragment is

```
        ...
product equ   4321H               ; allocate PRODUCT the
                                  ; address 4321H
        ...
        mov   al,byte ptr 1234H   ; put the byte in
                                  ; 1234H into AL
        mov   bl,89               ; put the other byte
                                  ; factor in BL
        imul  bl                  ; multiply AL by BL
                                  ; getting the result in AX
        mov   product,ax          ; copy the result
                                  ; to PRODUCT
        ...
```

The instruction for multiplication has only one operand because one factor is always in a specified register. If bytes are being multiplied, one factor is in AL and the product is placed in AX. For the multiplication of words, one of the words has to be in AX. Since the product can be a double word the high order word of the answer is placed in DX and the low order word in AX.

2.8 Division

The register AX also occupies a special position in division. The dividend is in AL if a byte and in AX if a word. The divisor is not permitted to be an immediate value but must be in a register or memory. If the result of the division is in bytes the quotient goes in AL and the remainder goes in AH. For words the quotient appears in AX and the remainder in DX.

Suppose we want to divide a word located at 1234H by 89 and keep the 89 conveniently for future divisions. The quotient is to be located at 4321H and the remainder in BX. Then the program fragment could be

```
         ...
quotient equ   4321H              ; allocate QUOTIENT the
                                  ; address 4321H
```

```
...
mov     ax,word ptr 1234H  ; transfer the
                           ; dividend to AX
mov     cl,89              ; put the divisor in CL
idiv    cl                 ; divide AX by CL
mov     quotient,ax        ; copy AX to QUOTIENT
mov     bx,dx              ; copy the remainder
                           ; to BX
...
```

Notice that we have kept 89 in CL so that it is available for further divisions.

Positive and negative numbers

2.9 Working with fixed registers

In high level languages one tends to be fairly cavalier about arithmetical calculations unless one is manipulating very large or very small numbers because one expects things to be taken care of behind the scenes. Rather more attention has to be paid in assembly language, since the registers are of fixed size. Another difficulty arises from the fact that all representations are in binary notation so that somehow negative numbers must be indicated via 1 and 0 since the symbol − itself is not available.

Let us start by considering non-negative numbers. The jargon for these is *unsigned magnitudes*. For simplicity let us suppose that a register is limited to four bits instead of a byte. Then the register can display all binary numbers from 0000 to 1111 or from 0 to 15 in decimal. Four examples of addition and subtraction are shown below with the boundaries of the register as dashed lines:

addition		subtraction	
¦ 0011 ¦	¦ 1000 ¦	¦ 1011 ¦	¦ 0011 ¦
¦ 1001 ¦	¦ 1001 ¦	¦ 1001 ¦	¦ 1001 ¦
¦ 1100 ¦	1 ¦ 0001 ¦	¦ 0010 ¦	−1 ¦ 1010 ¦

In two cases there are no problems but for one addition the sum exceeds the capacity of the register. The figures in the register are correct but incomplete because a fifth slot is needed. On the other hand in the troublesome subtraction an attempt is being made to create a negative number (namely −10000B + 1010B) when we have already agreed to stay with unsigned magnitudes only.

Whenever the addition of two numbers exceeds the capacity of a register an extra slot is needed for the 1 which should be carried. You will see in Section 4.2

that there is a special instruction on the 8086 which allows you to keep track of the carry by bringing into play a further register. Similarly, there is a special instruction for coping with a borrow on subtraction.

There are also difficulties with multiplication and division. If two 4 bit numbers are multiplied their product can occupy a byte; indeed the product can range from 0 to 225 in decimal. Therefore processors generally have provision for registers to cope with double the length of the factors as has been seen in Section 2.7.

Division can be regarded as the inverse of multiplication so that we might expect to be confronted with a dividend of double the length of the divisor. An example, in decimal since most people do not take kindly to binary division, is 3976÷49 giving a quotient of 81 and a remainder of 7. Both quotient and remainder fit into the same size register as the divisor.

If we had been asked for 76÷49 the dividend does not apparently have four digits expected of a double length dividend. That can be overcome by adding two zeros to make it 0076. Again the quotient (1) and the remainder (27) fit the same size register as the divisor.

However, 3976 ÷ 39 is not straightforward. The quotient will require three digits and, therefore, a bigger register than the divisor. If this is not provided there will be an overflow problem in division.

2.10 Signed numbers

Unsigned magnitudes may be satisfactory for some purposes such as ASCII codes but in general we cannot manage without negative numbers. Since we do not have a minus sign we are going to have to devise a code which identifies negative numbers. A byte can take 256 values; consider a code in which 0 to 127 are unaltered but those from 128 to 255 are to represent negative numbers in some way. The reason for trying this is that the most significant bit of 0 to 127 is 0 whereas that of 128 to 255 is 1 so that negative numbers would be readily recognized. Now, if we take 256 from the range 128 to 255 we obtain the range −128 to −1 which just tags on nicely to our positive range of 0 to 127. This is the code that we are going to adopt for *signed numbers*, namely that 0 to 127 represent themselves whereas 128 to 255 are codes for −128 to −1. A table showing the binary codes of positive and negative numbers side by side is in Fig. 2.4.

Decimal	Binary	Binary	Decimal
+127	0111 1111	1000 0001	−127
+ 98	0110 0010	1001 1110	− 98
+ 32	0010 0000	1110 0000	− 32
+ 3	0000 0011	1111 1101	− 3
+ 1	0000 0001	1111 1111	− 1
0	0000 0000	0000 0000	0

Fig. 2.4 Signed numbers

The code for the negative number is derived by adding 256 or 2^8 to the negative number. In general, negative numbers are coded by adding a power of 2 (2^{16} for words) and so signed numbers are often known as *twos-complement numbers.*

From Fig. 2.4 you can see that a rule for getting the binary representation of a negative number from its positive counterpart is to change 1 to 0 and 0 to 1; then add 1 to the most right-hand bit. You should note, however, the seeming anomaly that the twos-complement of -128 is -128. The explanation is that -128 has no positive counterpart. Another way of deriving the binary form of a negative number from the positive is to begin at the left and change 0 to 1 and 1 to 0 until you reach the last 1 on the right; do nothing with that and make no further changes.

As has already been pointed out negative numbers are distinguished from positive by having a 1 in the most significant bit instead of a 0. For this reason the leftmost bit of a signed number is called the *sign bit.* However, in arithmetical operations the processor treats it the same as any other bit. This can lead to complications in addition and subtraction just as for unsigned magnitudes.

For example, consider 4 bit signed numbers. They can represent numbers in the range from -8 to 7. Therefore if we add 3 and 4 there is no problem but if we try 7 and 4 we get

```
 0111
 0100
 ----
 1011
```

i.e. -5. The correct answer cannot be represented in the four bit register and so there is overflow. Again -7 plus -4 would be outside the register range and there would be overflow. The actual calculation together with that of -3 plus -4 is

```
 | 1001 |        | 1101 |
 | 1100 |        | 1100 |
 | ---- |        | ---- |
1| 0101 |       1| 1001 |
```

In the first case the number in the register is positive, whereas in the second case the content of the register is the correct -7; the carry is irrelevant.

Subtraction suffers from similar difficulties. Often the hardware is arranged so that instead of subtracting a number the twos-complement is added. From our point of view this is an equivalent procedure and has no influence on the programming.

The presence of a sign bit obviously influences how multiplication and division should be undertaken. Consequently there are two multiplication instructions, MUL for unsigned magnitudes and IMUL for signed numbers or twos-complements. Similarly, DIV is for the division of unsigned magnitudes whereas IDIV is for signed numbers.

There is another special instruction associated with division. Given a byte dividend in AL you may wish to regard it as a word in AX for the purposes of division. The byte must then be extended in a suitable way. For unsigned magnitudes there is no problem, because the number is never negative—just fill AH with zeros. The same procedure is satisfactory for a positive signed number. However, if the byte in AL is negative, the sign bit in AL is 1 and completing AH with zeros will result in the word AX having a sign bit of 0 i.e. the extended number will be of the opposite sign to the original. What has to be done is that AH is filled with 1s. The instruction CBW takes care of these variations for signed numbers, making AH all zeros or all ones according as the sign bit of the byte in AL is 0 or 1. In fact, there is also an instruction CWD which converts a signed word in AX into a double word, the extension occupying DX.

Methods for discovering if overflow etc. has occurred depend on knowledge of the following sections.

Flags and jumps

2.11 More on division

As a further illustration of the differences let us consider the division of the byte in AL by the byte in CL with the quotient left in AL, the remainder in location REM and AH cleared.

The general rule for processors is that they divide words by bytes rather than bytes by bytes. So, firstly, the dividend is extended to a word. Suppose that we are dealing with signed numbers. Then we have

```
...
cbw                 ; extend byte in AL to
                    ; word in AX with sign
idiv    cl          ; divide AX by CL, getting
                    ; quotient in AL, remainder in AH
mov     rem,ah      ; transfer remainder to REM
xor     ah,ah       ; clear AH
...
```

Notice the division instruction for signed numbers and the way in which AH is cleared. No instruction has the specific task of clearing a register and so others have to be adapted. The XOR instruction combines corresponding bits of the source and destination according to the logical rules for XOR in Section 1.5(b). The given instruction therefore results in AH being 0. At least two other methods exist for clearing, namely SUB AH,AH and MOV AH,0 though MOV tends to be slower than the other two.

For unsigned magnitudes CBW is inappropriate for extension since we always want AH full of zeros. In the case we have

```
        . . .
        sub     ah,ah       ; clear AH
        div     cl          ; divide AX by CL, getting
                            ; quotient in AL, remainder in AH
        mov     rem,ah      ; transfer remainder to REM
        xor     ah,ah       ; clear AH
        . . .
```

where, for the sake of illustration, two different ways of clearing AH have been deployed.

2.12 Flags

A description of the flags and how they are affected by the most recent calculations has already been given in Section 2.3. Now that we know more about operations with numbers it is convenient to amplify that description.

Firstly, the sign flag (SF), which indicates that the latest computation has resulted in negative number, is actually raised to 1 when the most significant bit is a 1. This means a negative number only for signed numbers; for unsigned magnitudes it can be ignored unless you want to detect whether the most significant bit is a 1.

The carry flag (CF) is set to 1 if there is a carry from addition or a borrow from subtraction in the highest order bit. With unsigned magnitudes this means that the result either overflowed or was a negative number; neither possibility can be coped with by unsigned magnitudes in a fixed register size. If you are working with bytes you may wish to consider converting to words of unsigned magnitudes or signed numbers depending on context.

The overflow flag (OF) is primarily for use with signed numbers; it has no significance for operations with unsigned magnitudes. When it is raised to 1 it conveys the information that the result is outside the capacity of the register for signed numbers which generally signifies that there has been overflow into the sign bit.

While the state of the flags tells you whether something untoward has occurred nothing is done to remove the source of the effect. Proper planning by you is essential especially in connection with division. When the quotient will not fit into the space allocated to it, as when you try to divide by 0, the operating system is likely to stop the program and display the message 'Divide overflow'. Obviously, you should provide a safeguard in your source program to prevent division by 0 but, as a matter of good practice, you should make some rough calculations to check that any quotients expected to arise will not exceed the appropriate ranges for unsigned magnitudes and signed numbers.

2.13 Jumps

The 8086 has essentially one *unconditional jump* instruction JMP. It forces the processor to transfer to another part of the program and is therefore useful for leaping over a set of instructions e.g.

```
          ...
          mov     bx,ax
          jmp     sum
          ...
sum:      ...
```

in which the program, after copying AX to BX, is obliged to follow the instructions labelled SUM. A leap forward is shown but a backward jump is equally feasible. Some assemblers have special instructions for a *far jump*, JMPF, which may involve transfer to another code segment, and for a *short jump*, JMPS, within ± 128 bytes but other assemblers cover both possibilities by JMP itself. The virtue (if any) of characterizing a short jump is that the assembler may be able to assemble the jump in a smaller number of bytes. Short jumps are normally restricted to those in the forward direction so as to advise the assembler whether a short or far jump is appropriate; for a backward jump the assembler knows what the separation is.

All *conditional jumps* are short jumps to labels which must be within ± 128 bytes. If you want to go further afield with a conditional jump you have to do it by programming; for example, make a legitimate conditional jump to an unconditional jump pointing to the desired destination. It is not easy to lay down a precise rule which ensures that a conditional jump lies within the range ± 128 bytes because the object code of a machine instruction may occupy anything from one to six bytes. A rough criterion is that, if the target label is not more than 25–30 instructions from the jump, it can be assumed that a short jump is involved. Should you make a mistake the assembler will report an error such as 'Label out of range', indicating that the target is more than 128 bytes from the jump instruction.

Conditional jumps are of the form J· where the dot stands for one, two or three letters. These letters are mnemonics to remind you what is being tested in the result to decide whether or not a jump is made. For instance, JZ means 'jump if the result is zero' whereas JNZ means 'jump if the result is not zero'. The processor decides whether or not to jump by examining the flags. Thus JZ causes a jump if the zero flag is raised i.e. ZF = 1 while JNZ effects a jump if the zero flag is cleared i.e. ZF = 0. A complete list is given in Appendix C. Notice that some of the instructions have the same effect; you may find one or other more convenient in different contexts. Do not worry if the assembler does not reproduce your precise mnemonic but an equivalent form; the running of the program will not be affected.

In arithmetical operations we shall want to protect ourselves from the various faults that can arise with registers of fixed size. Conditional jumps form a convenient vehicle for checking errors. For example, suppose that we are working with unsigned magnitudes and plan to subtract BL from AL. If BL is greater than AL the correct answer is negative which is not an unsigned magnitude. A borrow would be involved. You might think that AH could assist but AH and AL are independent registers unless combined into AX. The borrow would set the carry flag. So we check it; if it is down we continue with operations but if it is raised we go to other instructions which might, for instance, print an error message. A program fragment might be

```
            ...
            sub    al,bl      ; form AL - BL in AL
            jc     crry       ; check the carry flag: if
                              ; CF = 1 jump to label crry
            ...               ; otherwise continue
crry:       ...
```

Note the colon on CRRY in the last line, showing that it is a label for a set of instructions starting there.

If AL and BL are signed numbers the problem is slightly different because now overflow can occur e.g. if AL is positive and BL is negative the result might exceed the capacity of the register. So our fragment for signed numbers would be

```
            ...
            sub    al,bl      ; form AL - BL in AL
            jo     ovrflw     ; check the overflow flag
                              ; if OF = 1 jump to ovrflw
                              ; otherwise continue

            ...
ovrflw:     ...
```

2.14 Comparison

A standard route entailing a conditional jump is a comparison procedure where the path goes one way or another depending on the result of the comparison. A typical situation is where you want to finish if AL is 0FFH but otherwise continue the process by, say, clearing BH. A suitable fragment is

```
        ...
        cmp    al,0ffh    ; compare by forming AL - 0FFH
        je     finish     ; if AL = 0FFH jump
        xor    bh,bh      ; otherwise clear BH
        ...               ; and continue
finish: ...
```

Observe that the operation of CMP has no effect on AL. After CMP has completed its subtraction the flags are adjusted, but the result is not stored in any other manner.

A slight variant on the above is to examine whether CX is less than 0FFFH.

If not, jump to FINISH; if it is, increase CX by 1 and continue. This leads to

```
        ...
        cmp     cx,0fffh    ; compare by forming CX - 0FFFH
        jnl     finish      ; if CX not less than 0FFFH jump
        inc     cx          ; otherwise add 1 to CX
        ...                 ; and continue
finish: ...
```

For comparison of quantities which are not in registers a transfer is necessary first. Suppose we want to jump if FIRST > SECOND, both words being signed numbers, but not otherwise. Then we proceed as follows:

```
        ...
        mov     ax,first        ; copy first word to
                                ; working register
        mov     bx,second       ; copy second word to
                                ; working register
        cmp     ax,bx           ; compare by AX-BX
        jg      over            ; if AX > BX jump
                                ; otherwise continue
        ...
over:   ...
```

If it had been the content of the location FIRST rather than its value that concerned us then, in the MOV instruction, it would have been necessary to use WORD PTR FIRST.

Somewhat more complicated jumps may involve more thought to achieve economy in assembly. Suppose, for example, if AX > BX we want to place AX in CX but otherwise put BX in CX. One method might be

```
        ...
        mov     cx,ax           ; transfer AX to CX
                                ; in case AX > BX
        cmp     ax,bx           ; compare by AX - BX
        jg      keep_on         ; if AX > BX keep going
                                ; since CX has been adjusted
        mov     cx,bx           ; otherwise copy BX
                                ; to CX
keep_on: ...                    ; continue program
```

Observe that, whether AX is greater than BX or not, the program continues with the instructions labelled KEEP_ON. May I remind you that KEEP_ON is not the same label as KEEPON.

The approach following the conditions more directly is

```
        ...
        cmp     ax,bx           ; compare by AX - BX
```

```
          jg      move_ax     ; if AX greater than
                              ; BX, jump
          mov     cx,bx       ; otherwise put BX in CX
          jmp     keep_on     ; leap over MOVE_AX
move_ax:  mov     cx,ax       ; put AX in CX since AX>BX
keep_on:  . . .               ; continue program
```

This program contains an extra jump, an unconditional one, in order that the adjustment to CX for AX < BX is preserved and not modified by the instruction appropriate to AX > BX. Although the second version follows more closely what we are asked to do the first is shorter and cleaner. Do not feel obliged to tread the precise footsteps of a formulation if there is a better equivalent route.

Exercises

1. What are the largest and smallest decimal numbers that can be represented in a data register using (a) unsigned magnitudes, (b) signed numbers?
2. Which of the following are legal labels?

```
nolist     First_street     How_are_you     this week
AH         1st_street       :colon          dash$flag
8086       Firststreet      colon:          number9
```

3. Which of the following lines of code are legitimate?

```
seven      equ     7
movv       equ     mov
clear:     xor     ah,ah       clear AH
sum        add     bh,cl     ; find sum
           and     ch,10
           and     mary,john
           add     dollars, pounds
;          sub     income,cost
           mov     house,jane
           mov     house,20
; jack:    jim;    joe!
           div     number,25
     William and Mary
```

In the following questions, write assembly language for the stated requirement

4. Add decimal 732 to BX.
5. Subtract the byte in memory location 1234H from the byte in memory location 4321H and place the result in location 5555H.
6. Reserve two bytes in memory. Place the byte from CH in the first and that from CL in the second.

7. Add the byte in memory location FIRST to the byte in AL and place the result in FIRST.
8. The product of the words in FIRST and SECOND occupies one word. Place it in THIRD.
9. FIRST contains an unsigned magnitude word. Multiply it by 9876 and place the highest 2 bytes in CX. Divide the word in FIRST by 35 and put the quotient in BP, the remainder in SI.
10. Add the word in BX to AX. If there is an overflow jump to the label BAD, otherwise jump to the label GOOD.
11. AL and BL contain unsigned magnitudes. (a) If BL > AL jump to BAD otherwise continue. (b) If BL > AL continue, otherwise jump to BAD.
12. AX, BX, and CX contain unsigned magnitudes. Move the larger of AX and BX to DX and then add the largest of AX, BX and CX to DX.
13. The byte in AL represents an ASCII character. If it is a lower case letter change it to the corresponding capital. If it is a capital continue in sequence, otherwise clear BX.

3

CYCLING

3.1 Repetition

You will be aware already from your experience with high-level languages that the strength of a computer lies in its ability to repeat at high speed a sequence of instructions as many times as you wish. The machine thus performs, as often as you ask it, a *cycle* (also called a *loop*, though some reserve the word loop for a cycle which the machine cannot leave unless it is switched off; further, LOOP is also reserved as a key word for the 8086).

A simple cycle in BASIC is

```
10  FOR I=1 TO J
20  A=A+I
30  NEXT I
40  END
```

The condition in 10 tells the machine how many circuits to undertake. At 30 the computer checks if I is less than J. If it is, I is increased by 1 and the computer goes back to 20. If, on the other hand, I = J at 30 the computer proceeds to 40. Since the test is at the end the computer must go through the calculation at least once and this is always true of a FOR sequence. If you desire the possibility of never doing the calculation then there must be a test at the beginning of the cycle or, in the case of BASIC, an earlier instruction such as

```
5  IF  J=0  THEN 40
```

Another point is worth noticing. In the BASIC program above the machine does the counting without any conscious effort on your part. In assembly language you will generally have to provide a counting mechanism explicitly, except for one or two instructions which do some counting automatically. The following is an illustration where the machine cycles as many times as the positive number in CX*

```
            . . .
count:      cmp    cx,0       ; test whether CX=0
            je     over       ; if CX = 0 jump to OVER
            dec    cx         ; decrease CX by 1
            jmp    count      ; go round again
over:       . . .             ; continue program.
```

The instruction DEC diminishes the number in CX by 1. So, on each circuit, CX is reduced and the process continues until CX is zero. Notice that the test for this is set right at the beginning so that if CX is zero to start with no cycling takes place. The program has a flaw in that if, by accident, CX is initially negative there is no way of leaving the cycle. This defect can be circumvented by replacing JE with JLE though, in a more complicated program, it may be desirable to have a separate check that the initial value of CX is in the specified range. In general, programming in assembly language entails much more attention to testing that errors in one part of the program do not cause unexpected effects when another part is entered.

Actually, DEC has another interesting property. It adjusts the flags after the decrement. Thus, if the result is zero the ZF flag is raised to 1 but is cleared if the result is non-zero. Therefore, instead of jumping back to count, we could go back to the instruction JE without any loss. In fact, the adjustment of flags enables the shortening of the code (which can be important when memory space is at a premium)

```
         ...
count:   dec    cx       ; decrease CX by 1
         jnz    count    ; if not zero, repeat
         ...             ; continue
```

The basic counting method described can be used in conjunction with other instructions. Watch out that these do not affect CX or, if they do, that an appropriate recovery mechanism is provided. Suppose that the cycle is to be undertaken 1000H times. Then we might have

```
        ...
        mov   cx,1000h  ; place count value
                        ; in CX
        cmp   cx,0      ; check whether counting
                        ; to be started
cycle:  jz    over      ; test to see if CX is
                        ; zero
        ...             ; instructions to be repeated
        dec   cx        ; reduce count by 1
        jmp   cycle     ; round again
over:   ...             ; continue program
```

If there are numerous intermediate instructions there is another pitfall to look out for, namely that JZ cannot take you more than 128 bytes. When the distance is too large you may have to introduce subsidiary jumps or restructure the program.

3.2 Arrays

Repeated addition is not an uncommon feature of calculations. It is therefore an obvious candidate for the construction of a cycle. However, there may be occasions when this is not so. When there are not many numbers it may be faster to add each one separately to a register, such as AX, so that the sum accumulates there. In this way, the time spent on the instructions concerned solely with arranging a cycle is saved. Again, when the numbers are scattered about the memory the construction of a convenient circuit may not be easy.

Of course, scattering the numbers in memory is to be avoided as far as possible. Hence, we usually try to arrange them in an *array* of consecutive memory locations. Arrays can be handled in high level languages simply by declaring the range of subscripts to be employed. This avenue is not open in assembly language because of the absence of subscripts; the programmer has to arrange the necessary change of address from one number to the next.

Suppose that there are ten numbers, each a word long, to be placed in consecutive locations under the label ARRY. We can reserve space for them by the assembler directive

```
arry    rw    10    ; reserve 10 words for ARRY
```

This tells the assembler to keep ten consecutive words with unspecified initial values on one side for occupancy; the assembler has the choice of location for the necessary 20 bytes of storage. Reservation of the storage for ARRY by RW means that ARRY has been given the *attribute* of WORD i.e. it is expected that 2 bytes will be manipulated when reference is made to ARRY. Therefore an instruction such as

```
mov   arry,cl
```

will cause the assembler to issue an error signal because you are attempting to move a byte to a destination designated for words. However, you can override the attribute by means of PTR. Thus

```
mov   byte ptr arry,cl
```

will transfer the content of CL to the first byte of ARRY. Similarly, a BYTE attribute can be overridden by employing WORD PTR.

Another feature of reserving space should be emphasized. Initial values can be ascribed to memory locations by the declaration DB. For example,

```
mem   db   1,2,3,4,5
```

will allocate five bytes to the label MEM giving them the values 1,2,3,4,5 in that order. The assembler always makes its provision sequentially. So, if we write

```
mem    db    1,2,3,4,5
       rs    12
```

the assembler will put aside 17 consecutive bytes. The first five bytes will be filled in with the values 1,2,3,4,5. The subsequent 12 bytes will be reserved, but with unspecified initial values.

Now, what is the significance of the label ARRY which has been attached to the array? In fact, ARRY is an *address value* i.e. the address of the first number in the array. Suppose that the assembler has allocated the address 1234H to the first element of the array. The first word will then occupy the bytes at 1234H and 1235H; the second word will be in 1236H and 1237H and so on. ARRY will have the address value 1234H and, whenever ARRY appears in the program, the assembler will replace it by 1234H.

The last statement merits some amplification. Since ARRY has been declared as an address value the assembler knows that it is not an immediate value. Therefore it will interpret

```
mov   ax,arry
```

as copy the content of ARRY to AX i.e. the first word of the array is copied to the register AX. To insert the address of ARRY in AX a special instruction LEA (load effective address) is issued e.g.

```
lea   ax,arry
```

The distinction between reserving space and using EQU is perhaps worth emphasising. Even if it was known in advance that the assembler was going to allot 1234H to ARRY we would not wish to open with

```
arry   equ 1234H
```

Firstly, it does not supply an array (additional directives would be necessary to do that) and, of course, WORD PTR would have to be included in instructions to copy content. Secondly, it would be a serious constraint on the assembler because, if the program were modified subsequently or combined with another, an alternative location for ARRY might be preferable. Generally speaking, the assembler should have as much flexibility and economy as feasible.

So far only the first word of the array has an address. Other words are addressed by adding on the requisite *number of bytes* from the first word. So ARRY + 2 is the address of the second word, ARRY + 4 that of the third word, and so on.

Indirect addressing is also available. Thus, if the content of BX is 2, ARRY[BX] points to the same address as ARRY + 2.

To work along an array a *pointer*, capable of giving the address of every element of the array, has to be generated. For easy change during the program the pointer should reside in a register and, since the pointer is effectively acting as an index, a suitable choice is SI.

Let us now consider the addition of ten words in an array. The code for a program fragment is

```
                . . .
        arry    rw      10          ; reserve 10 words for the array
        sum     rw       1          ; reserve 1 word for the sum
```

```
count   dw    10          ; declare COUNT as a word
                          ; which has the value 10
        . . .             ; program for inserting the
                          ; words in the array
        lea   si,arry     ; put the address of ARRY
                          ; in SI
        xor   ax,ax       ; clear AX for addition
        mov   cx,count    ; prepare CX for counting
        cmp   cx,0        ; test to see if CX is zero
cycle:  je    exit        ; if CX=0 the addition is
                          ; finished
        add   ax,[si]     ; add to AX the word which
                          ; SI is pointing to
        add   si,2        ; adjust SI so that it is
                          ; pointing to the next word
        dec   cx          ; reduce count by 1 (with
                          ; flag adjustment)
        jmp   cycle       ; go back to test again
exit:   mov   sum,ax      ; place the result in SUM
        . . .
```

The strategy is that the numbers are added one-by-one and totalled in AX (having first cleared it to zero). The numbers are picked by the pointer SI which, therefore, has to be increased by two on each circuit; indirect addressing is used. CX keeps an eye on how many numbers have been added and so has to be decreased by 1 on each circuit (alternatively, CX could have been started at 1 and increased until a test revealed that it had reached 10). Note that, although DEC adjusts the zero flag, MOV does not (nor does JMP), so that the CMP has to be introduced to test whether any cycle is required.

3.3 Maximum of array

A standard facility in BASIC is finding the maximum of an array and it is of interest to see how this is accomplished in assembly language. The problem to be discussed is discovering the largest *absolute value* or *modulus* i.e. the signs of all the negative numbers are reversed while the positive numbers are unaffected and we seek the largest of this sequence of positive numbers. Clearly, an instruction for an appropriate change of sign will be required. However, if it is known in advance that the numbers are unsigned magnitudes, such an instruction will have to be dropped to avoid comitting an error because of the way in which negative numbers are represented.

Instead of fixing the length of the array at 10 as in the previous section we will allow the latitude of giving it the value NUMB which can then be chosen at your convenience. SI will still be used as the pointer to the address but another register will have to be brought into play since all manipulation cannot be carried out in AX. Indeed, each number will be drawn successively from the array to see whether its absolute value is greater than the largest one already found. On the basis that the numbers are word-size the program fragment will look like

```
                . . .
arrylen     equ     numb        ; specify the length of
                                ; the array by giving the
                                ; value NUMB
arry        rw      arrylen     ; reserve the right number
                                ; of words for the array
max         rw      1           ; reserve 1 word for
                                ; the maximum modulus
            . . .               ; program for placing
                                ; the numbers in the array
            lea     si,arry     ; put the pointer to
                                ; the array in SI
            sub     bx,bx       ; clear BX as home for
                                ; maximum
            mov     cx,arrylen  ; prepare CX for counting
                                ; to end of array
            cmp     cx,0        ; test if CX is zero
cycle:      je      exit        ; if CX=0 the maximum
                                ; has been found
            mov     ax,[si]     ; copy element of array
                                ; to AX
            cmp     ax,0        ; test for sign of element
            jge     pos         ; if positive do not change
            neg     ax          ; if negative change the sign
pos:        cmp     ax,bx       ; see if AX has larger value
                                ; than the one retained
            jng     same        ; if not greater maximum
                                ; is unchanged
            mov     bx,ax       ; otherwise put new
                                ; maximum in BX
same:       add     si,2        ; adjust pointer
            dec     cx          ; reduce count by 1
            jmp     cycle       ; go back to test again
exit:       mov     max,bx      ; place the maximum in MAX
            . . .
```

The basic strategy for the registers is that BX holds the current maximum, successive numbers (pointed to by SI) are placed in AX to see if they are larger and CX does the counting. So we start off by pointing SI to the beginning of the array, clearing BX (since the maximum modulus cannot be less than zero) and setting CX to the length of the array. This is followed immediately by a test to check whether CX is zero and no further cycling is necessary. Note that the last instruction of the cycle before jumping back to repeat it reduces CX by 1.

In the cycle the next number of the array is copied to AX means of indirect addressing via SI; SI is increased by 2 bytes (1 word) near the end of the cycle.

Next, the sign of the number in AX is checked. If the sign is not negative the subsequent instruction is skipped and the program moves to the instruction labelled POS. If the sign is negative the instruction NEG is applied which changes the sign of the number by subtracting it from zero.

The program then checks if the number is greater than the maximum stored in BX. If it is, it is placed in BX and becomes the new maximum. Otherwise, BX is left unchanged and the program jumps to the final instructions of the cycle.

3.4 Double cycling

As an illustration of cycling within a cycle we consider ordering an array of signed numbers so that they steadily increase from beginning to end. As before we assume that the array consists of words and that the length is specified in advance. The technique that we shall adopt is to compare two consecutive elements of the array and arrange that the larger one occupies the slot nearer the end. After the first traverse of the array the largest number is in the endmost position. The second traverse places the second largest number next to the endmost. Of course, there is one comparison fewer to be made on this traverse since we know that the largest number already occupies its rightful place. Keep on traversing the array until only the first two elements have to be compared and reordered if necessary. This mechanism for ordering an array is often known as a *bubblesort*.

A possible program fragment is

```
            ...
arrylen     equ     numb          ; specify length of array by NUMB
arry        rw      arrylen       ; reserve right number
                                  ; of words for array
            ...                   ; place numbers in array
            mov     bp,arrylen    ; set BP to length of array
begin:      dec     bp            ; decrease BP by 1 to give
                                  ; number of traverses
            je      over          ; if BP=0 no more traverses
            lea     si,arry       ; make SI point to the array
            mov     cx,1          ; start count of number
                                  ; of comparisons
repeat:     mov     ax,[si]       ; copy element of array to AX
            mov     bx,[si]+2     ; copy next element of
                                  ; array to BX
```

```
            cmp    ax,bx       ; if AX≤BX leave the order
            jle    no_change   ; of the two elements unaltered
            mov    [si],bx     ; otherwise interchange
            mov    [si]+2,ax   ; their order
no_change:  inc    cx          ; adjust the counter
            cmp    bp,cx       ; if BP<CX the traverse is
            jb     begin       ; finished so start next one
            add    si,2        ; otherwise, adjust SI for
            jmp    repeat      ; the next pair of words
over:                          ; continue with program
            . . .
```

You will observe that three registers BP, CX, and SI are involved in the counting process. BP keeps track of the number of traverses, CX looks after the number of comparisons within a traverse, and SI takes care of the addresses of the numbers being compared.

The operand [SI]+2 ensures an address of 2 bytes or 1 word from [SI] and so points to the next element of the array. An alternative way of writing it is 2[SI] with the offset ahead of and adjacent to the bracket. If the offset had been negative either [SI]−2 or −2[SI] would have been acceptable.

In the above fragment CX has been counted upwards until it reaches BP. An alternative route would be to start CX at the same value as BP and count downwards. Which do you think produces the better code?

Actually, there is a defect in the above code because, once a traverse has been reached in which the elements are correctly ordered, any further traverses are superfluous. Therefore, in the interests of efficiency, some instructions should be inserted to provide an exit as soon as correct ordering has been achieved.

3.5 Double arrays

Although multi-indexed arrays are permitted in high-level languages they are not so easy to arrange at the assembler level. Usually, it is accomplished by labelling the arrays or by placing subarrays end-on to form a single array (see also Section 4.7). Suppose n estimates $x_1, \ldots, x_n$ of some data are made by one person, whereas another person records the estimates $y_1, \ldots, y_n$ for the same data. Then

$$\frac{\sum_{i=1}^{n} |x_i - y_i|^2}{n^2 - n}$$

might be regarded as a measure of the correlation between the two sets of observations (it is assumed that $n > 1$).

Assuming that the estimates are bytes the x-values are to be stored at 1234H and the y-values at 4321H. The length of each array or n is a byte to be found at

8888H. The result of the calculation is to be placed at 888AH.

Two registers SI, DI will be needed to watch over the addresses of the arrays. CX will act as counter and BX will carry the running total. Any other manipulations will be carried out in AX, DX.

This time our program segment will set set out rather more fully:

```
; ---------------------------- Begin Declarations ----------------------------
;
                dseg                    ; establish a data segment
;
     arry1      equu    1234h           ; give starting address
;                                         of first array
     arry2      equ     4321h           ; give starting address
;                                         of second array
     arrylen    equ     8888h           ; address of array length
     result     equ     888Ah           ; address for result
;
;
;
; --------------------------- End of data segment ---------------------------
;
;
; ---------------------------- Initial statements ----------------------------
;
                cseg                    ; establish a code segment
;
     start:     lea     si,arry1        ; SI points to the first array
                lea     di,arry2        ; DI points to second array
                xor     ch,ch           ; set up CX as counter
                mov     cl, byte ptr arrylen
                xor     bx,bx           ; clear BX for running total
;
;
; --------------------------- End of initialization ---------------------------
;
;
; ------------------------------- Summation -------------------------------
;
     sumcycle:  jcxz    sum_over        ; if CX=0, summation finished
                xor     ah,ah           ; clear AH
                mov     al,[si]         ; put x in AL
                mov     dl,[di]         ; put y in DL
                sub     al,dl           ; form x-y in AL
                imul    al              ; AX=(x-y)^2
```

```
            add    bx,ax      ; add AX to running total
            inc    si         ; prepare for next
            inc    di         ; members of array
            dec    cx         ; adjust counter
            jmp    sumcycle   ; go round again
;
;
; ------------------------ End of summation cyle ------------------------
;
;
; ---------------------------- Final calculation ----------------------------
;
sum_over:   xor    ah,ah      ; clear AH
            mov    al,byte ptr arrylen
                              ; AX=n
            sub    cx,ax      ; CX= -n
            mul    al         ; AX=n^2
            add    cx,ax      ; CX = n^2 - n
            mov    ax,bx      ; place sum in AX
            xor    dx,dx      ; extend AX to double
;                             ; word for division by word
            div    cx
            mov    result,ax  ; copy answer to RESULT
;
; -------------------------- Calculation complete --------------------------
            end
```

There are several features to note about this layout. Although perhaps a little over elaborate for a short program it incorporates principles which are good practice when dealing with large programs. The primary aim is legibility and ease of detecting what is going on. The instructions have therefore been grouped logically and each group has been supplied with a descriptor so that you know what its purpose is. This is assisted by a liberal dose of comments. The semi-colon prevents the assembler processing this explanatory material so that it appears only in the listing of the program.

The declarations are all set out at the beginning with comments on their purpose. Some people prefer to have the declarations at the end (one reason being that in the construction of a program you often find that you want to add declarations), but this is a matter of taste. The names, as well as the labels for the groups of instructions, have been chosen to give some clue to their intent.

No doubt you can think of ways of improving the layout. For instance, you might wish to include the paragraph explaining how the registers were to be used in the listing. To save space subsequently we shall not always follow the pattern above but that does not in any way vitiate the statement that it is sound to get into the habit of logically breaking down a program into divisions which are easily comprehensible.

The program contains one new machine instruction JCXZ and three new assembler directives DSEG, CSEG, and END. Assembler directives tell the assembler what to do; they do not affect the execution of the program by the machine.

The instruction JCXZ tests CX. A jump occurs only when CX = 0. The only restriction is that the jump must be within ±128 bytes.

CSEG and DSEG, abbreviations for Code Segment and Data Segment respectively, are necessary because every instruction and variable in a program must be in some segment. Machine instructions must be assigned to the Code Segment but directives can be placed in any segment. The segments can be alloted names, e.g.

```
data1    dseg
code2    cseg
```

If you do not specify names the assembler will assume that they are CODE and DATA. The advantage of names is that you can use them to separate segments of long programs physically for reasons of distinction or convenience of layout. The assembler will combine all segments with the same name. Apart from this, the assembler assigns all statements after a segment directive to the specified segment until it encounters another segment directive.

Segment directives allow you to divide your assembly language source program into segments that correspond to the memory segments into which the resulting object code is eventually loaded when the program is run. How you arrange the segments determines how the operating system executes the program. Normally, you would have separate Code and Data Segments though it is acceptable, if not desirable, to mix the code and data in a single segment. In that case the combined code and data must not exceed 64K in length whereas, with distinct segments, the code and data can each be up to 64K long. Additional flexibility is provided by being able to specify ESEG and SSEG, the Extra and Stack Segments. The address space of the processor can be used more fully by creating a number of segments. In particular, you can have more than 64K of code or data by having several segments and managing them with assembler directives.

END tells the assembler that it has reached the end of program and that it can ignore any subsequent lines. The END directive can be omitted, though it is not good practice to do so; if it is omitted the assembler will continue processing until it finds an end-of-file character (1AH).

3.6 End-of-file

In the final paragraph of the preceding section it was observed that processing is continuous, in some circumstances, until an end-of-file character is encountered. Expressed another way, cycling is pursued until a special character indicates the need for change. Such an eventuality is not uncommon, the markers most frequently sought being 1AH (end-of-file), 0DH (carriage return) and 0AH (line feed). Therefore we examine how we might locate the end of a file, though the principles are equally applicable to the pinpointing of any particular character.

The problem is different from those met so far in this chapter. In the earlier examples the number of cycles to be undertaken was known in advance and the counter could be set at the beginning. Now we do not know how many cycles are involved, but have to carry on until we recognize the closing symbol.

Suppose that the array of bytes, of unknown length, starts at the memory location 1234H. Find the character 1AH and place its address in the memory location labelled FILE_END:

```
                dseg
        arry    equ     1234H       ; starting address of array
        eof     equ     1ah         ; identify character sought
        file_end rs     2           ; reserve a word for character
;                                     address
; ---------------------------------------------------------------
;
                cseg
                lea     si,arry     ; SI points to start
;                                     of array
        cycle:  cmp     [si],eof    ; check if byte indicated
;                                     by SI is end-of-file
                je      over        ; if so, leave cycle
                inc     si          ; if not, increase pointer
                jmp     cycle       ; and go round again
;
        over:   mov     file_end,si ; put address of 1AH
;                                     in proper memory location
                . . .
```

A flaw in the program is immediately evident. If, by some chance, the end-of-file marker is absent from the array there is no means of getting out of the cycle and it will be pursued indefinitely.

To see how this might be coped with consider summing positive words, in an array labelled ARRY, until either a zero is met or 1000 terms have been summed. On account of the large number of terms which may occur, the total may well occupy a doubleword.

```
                dseg
;
        terms   dw      1000        ; set the word at memory
;                                     location TERMS to the
```

```
;                                  value 1000
        total     rw     2         ; reserve two words for
;                                    the total
;
; ---------------------------------------------------------------------------------
;
                  cseg
                  xor    ax,ax     ; clear AX for the lower
;                                    word of the total
                  xor    dx,dx     ; clear DX for the upper
;                                    word of the total
                  lea    si,arry   ; SI points to first
;                                    word of array
                  mov    cx,terms  ; prepare counter
                  cmp    cx,0      ; check CX is not zero
        cycle:    je     over      ; if CX=0, finished
                  mov    bx,[si]   ; BX is next word of array
                  cmp    bx,0      ; test if word is zero
                  je     over      ; if it is, finished
                  add    ax,bx     ; otherwise, add word
;                                    to total
                  jnc    save      ; if no carry save DX
                  add    dx,1      ; otherwise add 1 to DX
        save:     add    si,2      ; adjust pointer by 2 bytes
;                                    for next word in array
                  dec    cx        ; reduce counter
                  jmp    cycle     ; go round again
;
        over:     mov    total,ax  ; put lower part of
;                                    doubleword in memory
                  mov    total+2,dx ; put upper part of
;                                    doubleword in memory
                  end
```

The cycle possesses a counter which limits the number of circuits to 1000. At the same time there is a test so that the cycle is left before the maximum number of terms if a zero should turn up.

Each word of the array is brought into BX. The running total is regarded as a doubleword, with the lower half in AX and the upper half in DX. The word in BX is added to that in AX (Remember: a word can be added to a word or a byte to a byte but not a word to a doubleword or a byte to a word). Then it is checked whether the result of the addition to AX is a number which exceeds the capacity of AX; if so, the carry flag CF will be set and an adjustment must be made to DX; if not, CF will be zero and DX can be left alone. On the other hand, if CF = 1 there is no jump and, since only positive words arise, a 1 must be added to DX to take account of the carry. If the words had both signs, extra manoeuvring would be required.

Notice that in the last section the Intel convention of placing the lower half of a quantity in the memory first has been followed.

3.7 Postscript on segments

Maybe a word of explanation about the convention for addresses would be helpful. The absolute address consists of two parts, the start of a segment and the offset. The start of the code segment, for example, is contained in CS. Our convention is that the segment registers are controlled by the operating system (though the programmer can override this when appropriate). Generally speaking, therefore, numerical addresses appearing in our programs are actually the offsets of the absolute addresses.

Normally, it is not a good idea to write programs with fixed memory locations. They lack the flexibility stemming from leaving it to the assembler. Moreover, the programs are not re-entrant. *Re-entrant code* can be interrupted and then re-entered by an interrupt service routine without loss of control or data as will be seen in a later chapter. Nevertheless you should know how to use fixed addresses in case you are in a situation where they are unavoidable.

Usually, the assembler locates the material in, say, a data segment starting from the beginning of the segment. However, the programmer can alter the origin by means of the ORG directive. Thus

```
dseg
org     100h
```

will set the address of the first data element so that it is offset 100H from the beginning part of the segment from the program being assembled. A collection of segments whose total length is less than 64K form a group which is addressable from a single segment register. Thus ORG permits the breaking up of a program into fragments of code, while allowing the fragments to be addressed without changing the contents of a segment register. This leads to more efficient code since it is faster and shorter than addressing fragments with 32-bit pointers.

Exercises

For exercises 2–5, 10, and 13 you should devise assembly programs for the stated requirement

1. If a (a) byte, (b) word, (c) doubleword starts at memory location 12FEH what is the address of the succeeding byte?
2. Ten words start at the memory location ARRY. Move them by a cycle to the location starting at NEW_ARRY.

3. The length of an array of bytes starting at ARRY is a byte in LEN. Find the number of bytes which are 1 and place the result in NUM.
4. An array of 60 words starts at ARRY. Sum every other word and place the total starting with the first word in ODD and that starting from the second word in EVEN. (Assume that the totals are words). How would you modify your program if the array were of bytes?
5. An array of 60 bytes consists of unsigned magnitudes. The sum of the elements which have odd values is to be placed in ODD and the sum of the remaining numbers in EVEN.
6. The successive bytes starting at ARRY are FFH, 00H, FEH, FFH, FFH, FFH, 85H, 02H, FDH, FFH, FDH, FEH, FCH, F4H, A2H, F1H, F6H, EFH, 5AH, 04H, FFH, FEH, 00H, 00H. If, in the example of Section 3.3, NUMB is 10 what is placed in MAX?
7. With the conditions of Exercise 6 what are the contents of ARRY after one traverse of the array by the program of Section 3.4?
8. Recast the code of Section 3.4 so that CX counts downwards.
9. Improve the code of Section 3.4 so as to prevent further traverses once the correct order has been accomplished.
10. The words starting at RECORD are signed numbers and there are 50 of them. The positive ones are added. If their total exceeds the capacity of 1 word change all the words to zero.
11. Modify the end-of-file program in Section 3.6 so that it gives the number of characters in the file in the word at LEN.
12. Modify the summation program in Section 3.6 so that the words can be signed numbers.
13. A list of 100 ASCII characters starts at FILE. Put the number of lower case letters in LCASE. (The codes of lower case letters extend from 61H to 7AH.)
14. SI points to a byte which is supposed to be maintained at 1111 1111 in binary. Due to a small fault it sometimes changes. Devise a program that will jump to the label ERROR to signal the occurrence of a fault.
15. In Exercise 14 the byte is supposed only to be negative. Construct a program that will jump to the label ERROR if it fails to be negative.
16. What is the difference between INC BL and ADD BL,1?

4

SUBROUTINES

4.1 Modules

You are probably familiar with subroutines in high-level languages and are aware of théir advantage in allow you to break down a program into smaller more convenient portions. In planning these portions you aim to keep the size down so that the writing and debugging are relatively straightforward. I am, of course, assuming that you adopt a *top-down* approach in which you have a clear view of the overall task and split it up into well-defined portions whose connections are carefully thought out in advance. The *bottom-up* strategy, in which you hope that bits of programs will grow like Topsy into a full-scale program, is apt to be frustrating when you attempt to tie the bits together.

Subroutines are blocks of code that can stand in their own right, accepting input and providing output. Thus EXP in BASIC is supplied with a value and returns its exponential. There may be other blocks which are independent of the rest, except for input and output, which you may not regard as subroutines and yet are worth constructing. Rather than referring to them as blocks, which tends to have a specialized meaning as in 'PASCAL is a block-structured language', we call them *modules*. To repeat then, a module is a set of code whose only connection with the remainder is via input and output. Here input and output are used in a broad sense so that statement is not meant to imply that a module cannot call upon another for assistance.

The essential feature of a module is that it contains a complete algorithm like the EXP of BASIC. As far as the program which calls upon it is concerned it looks like a black box. The calling program does not ask the module how it processes its task though it does expect the module to make the right calculation (if it does not, send for the module designer).

For this procedure to work there has to be an agreed form in which information is fed to the module. Equally well, the calling program has to accept the output in an agreed format. And, naturally, there must be some way of conveying suitable warning messages if the conventions are contravened. Therefore, to design a module requires full and precise information on the interface and what is to be accomplished internally.

The quantities which pass to and from the interface are known as *parameters*. This term is interpreted widely. For example, if there is a module for copying a file to another, the instruction might be

COPY FILEA FILEB

Then FILEA, the input to the module, and FILEB, the output from the module, would be called parameters.

The module may itself generate variables in fulfilling its purpose which are never seen in the outside world. They are known as *local variables*. In effect, the module picks up input parameters from specified spots, manipulates them and local variables, and then returns output to an agreed destination from which it can be retrieved by the calling program. The module, in its manipulations, may adjust registers which the calling program has already deployed. At the very least, the documentation of the module should warn the caller of registers that may be affected so that adequate protection may be incorporated in the calling program. Alternatively, the module should automatically preserve any registers which it modifies unless the possibility of such modification is known about in advance.

The registers form a natural habitat for the parameters which have to be passed to and from modules. It may be that you do not want to employ the registers for the parameters. In that case you have two options for where parameters may be deposited and collected from, namely memory locations or the stack. Examples of these different modes of passing parameters to and from an interface between modules will be given.

4.2 Example

Probably the simplest module that can be created is one for the addition of two numbers so that it is available for repeated summation. The first question is: is it for words or bytes? Words consume more register space but allow more scope for the summation. So we will settle on words. Where are the input words to go? Let us choose AX and BX. The output will go in DX so that the input does not have to be altered. Having fixed the locations of the parameters we can consider adding 1234H and 4321H by means of the module*

```
              ...
              mov     ax,1234H     ; put 1234H in AX
              mov     bx,4321H     ; put 4321H in BX
              call    sum          ; call module to
;                                    carry out addition
              ...                  ; continue with program
;                                    sum being in DX
      sum:    mov     dx,ax        ; copy AX to DX
;                                    preserving AX
              add     dx,bx        ; sum in DX, with
;                                    BX preserved
              ret                  ; return to calling
;                                    program
```

* All mnemonics copyright Intel Corporation 1986

The module is invoked by issuing the instruction CALL followed by the label of the module, namely SUM. The CALL instruction does not transfer any of the parameters; preceding instructions have to arrange that they are in the right places. What CALL does is first to put the value of the instruction pointer on the stack. Since IP is pointing to the next instruction this saves the position where the program is to return to after the module has performed its function. Once IP has been preserved the CALL instruction transfers control to the instructions bearing the label SUM. The module is terminated by RET. This unstacks IP so that the program can resume running at the instruction immediately after CALL.

A word of warning. You must ensure that the main program either ends before the module or jumps over it. Otherwise the main program will try to execute the module as part of itself with confusing consequences.

Before adopting SUM as a suitable general module it would be wise to include precautions which cover the possibility of capacity being exceeded. Since it is standard to place a doubleword in AX (lower word) and DX (upper word) it follows that the choice of AX for an input parameter is bad. So let us allot the input to SI and DI. Then we have

```
sum:    xor    dx,dx     ; clear DX
        mov    ax,si     ; first number in AX
        add    ax,di     ; add second number
        adc    dx,0      ; to AX and carry to DX
        ret              ; return to caller
```

The instruction ADC stands for 'add with carry'. It adds the contents of the source and destination, then adds CF. In this case the source is the immediate value zero so ADC merely adds the carry bit to DX.

Another variant would be to add the two numbers to the double word in AX, DX. Thus

```
sum3:   add    ax,si     ; add first value to AX
        adc    dx,0      ; and carry to DX
        add    ax,di     ; add second value to AX
        adc    dx,0      ; and carry to DX
        ret              ; return to caller
```

Instead of placing the words to be added in SI and DI we could use SI and DI to point to them. The only difference would be the replacement of si and di by [si] and [di] respectively.

4.3 Conversion

Suppose that the byte in AL is the equivalent of 103 in decimal and we wish to print this decimal equivalent. Usually, a printer is set up to accept ASCII characters. So, if the byte is sent direct to the printer, the printer will look up the ASCII table and then produce g. To avoid this hiatus the byte in AL must be converted into three bytes each one being the ASCII code for the corresponding digit of the decimal number.

The technique is to divide by 10 successively. After one division the quotient is 10 and the remainder 3, both expressed as binary numbers. The remainder supplies the last digit of the decimal but, before it can be stored, it must be converted to the ASCII representation by adding 48. Having dealt with the 3, we divide by 10 again. This time the quotient is 1 and the remainder is 0. Transform and store the 0. We may then either repeat the cycle or transform the 1 directly. The following module adopts the first alternative and assumes that the byte at BASE contains 10.

```
    destin     db     0,0,0           ; declare 3 consecutive
;                                       bytes to be zero
    conv_asc:  mov    bx,2            ; set up BX for byte
;                                       position
    cycle:     xor    ah,ah           ; clear AH
               div    base            ; divide by 10 remembering
;                                       that immediate value is
;                                       not allowed
               add    ah,48           ; make remainder into
;                                       ASCII equivalent
               mov    destin[bx],ah   ; put ASCII byte
;                                     ; in destination
               dec    bx              ; adjust byte position
               cmp    al,0            ; see if anything left
;                                       of binary number
               jne    cycle           ; and if so, go round again
               ret                    ; return to caller
```

DESTIN, here and subsequently, should be cleared inside the module in case the routine is called several times, when there would be the danger of earlier results interfering with later ones. The address DESTIN[BX] is interpreted as DESTIN + [BX]. Thus, when the content of BX is 2, it becomes DESTIN + 2. Consequently, by adjusting BX we can direct the three digits to the correct slots of DESTIN.

To direct the output to the next byte of DESTIN by means of DESTIN[BX + 1] or DESTIN[BX] + 1 is illegal. That is why BX must be adjusted before deploying the form DESTIN[BX].

Only the base registers BX, BP and the index registers SI, DI may be used in this fashion. It is permissible to combine a base register and an index register in square brackets so that, for example,

```
mov    ax,[bx+di]
```

is a legal instruction.

The above module assumes that unsigned magnitudes are being handled. For signed numbers there needs to be an extra byte in DESTIN to indicate the sign. The module might start

```
destin      db     '+', 0,0,0  ; declare sign and
;                                three bytes of zero
conv_ascs:  mov    bx,3        ; byte slot in BX
            cmp    al,0        ; see if AL is positive
            jge    cycle       ; if so, do division
            mov    destin,45   ; otherwise change sign byte
            neg    al          ; and reverse sign of
;                                number
cycle:      ...
```

and the instructions following CYCLE are the same as before.

The apostrophes in the declaration of DESTIN tell the assembler that the enclosure is a *character string*. The assembler therefore replaces the + by its ASCII code (43). A character string is not limited to a single character but its length must not exceed 255 bytes. The characters may be alphanumeric, special characters like brackets and punctuation, and spaces. The characters for carriage return, line feed, and tab cannot be inserted; they have to be provided separately if, say, a carriage return is required after the printing of a string.

Both lower-case and upper-case letters are acceptable in character strings; their form is retained by the assembler i.e. lower-case letters in a character string are *not* automatically translated into upper-case. Thus 'UPPER case' will be assembled into the ASCII equivalent of UPPER case. In order to include an apostrophe within a character string it must be entered twice. For example

```
'Size''s'   leads to   Size's
''''        leads to   '
```

A single character can be combined with a number, e.g.

```
alpha   db   'a'+80h
```

will result in 225 or hexadecimal E1 going to location ALPHA.

Declarations of character strings longer than 2 bytes are restricted to the directive DB. The directive for a word, DW, will accept strings of at most 2 characters and stores the low-order byte first. There is no provision in DD, the declaration of a doubleword, for character strings.

The apostrophe notation offers another route to conversion. Assume that unsigned magnitudes are involved.

```
destin      db     0,0,0       ; declare destination
```

```
table     db     '0123456789' ; declare character
;                              string of decimals
convasc:  mov    bp,2         ; BP indicates byte position
          xor    bx,bx        ; clear BX
cycle:    xor    ah,ah        ; clear AH
          div    base         ; divide by 10
          mov    bl,ah        ; remainder in BL
          mov    ah,table[bx] ; place right character
;                              from table in AH
          mov    destin[bp],ah; transfer character
;                              to destination
          dec    bp           ; adjust byte position
          cmp    al,0         ; number exhausted?
          jne    cycle        ; no, so round again
          ret                 ; back to caller
```

This module is both longer and more sophisticated than its predecessor. Yet it shows how, by modifying the BX register, you can pick out any desired character from a string and have it ready for storage in ASCII format.

So far the preservation of registers by a module has not been uppermost in our mind. Now we illustrate how this can be done by constructing a module for converting a binary number to hexadecimal. The binary number is assumed to be in the byte at VALUE and the hexadecimal code is to be placed in the two bytes at RESULT, the higher order byte occurring first. The strategy is the same as for conversion to decimal except that division is by 16 instead of 10. So the declaration BASE DB 16 will have to be available at some suitable site in the program. Since the result consists of only 2 bytes a cycle will be superfluous.

```
table     db     '0123456789ABCDEF'
hex_con:  push   ax             ; save AX on stack
          push   bx             ; save BX on stack
          mov    al,value       ; put number in AL
          xor    ah,ah          ; clear AH
          xor    bx,bx          ; clear BX
          div    base           ; divide by 16
          mov    bl,al          ; quotient to BL
          mov    al,table[bx]   ; hex character in AL
          mov    result,al      ; character to RESULT
          mov    bl,ah          ; remainder to BL
          mov    al,table[bx]   ; pick hex character
          mov    result + 1,al  ; character to destination
          pop    bx             ; recover BX from stack
          pop    ax             ; recover AX
          ret                   ; return to caller
```

The character string is shown separately because you may wish to declare it elsewhere for your convenience, but is juxtaposed to the module for convenience in following the instructions. The input parameter in VALUE and the output parameter in RESULT do not occupy registers, so that measures to retain quantities in the registers can be purely internal to the module. Also the arithmetic involves solely AX and BX so that attention can be confined to them.

The module starts by putting the words in AX and BX on the stack by means of PUSH and recovers them at the end via POP. They must come off in the opposite order to which they are stacked in order that the registers shall be unaltered when the module is left. Of course, you are not obliged to unstack an element in the same place that it started from if that is inappropriate. However, to prevent uninhibited growth or contraction of the stack every PUSH or POP must be associated with a POP or PUSH to keep the stack size on an even keel.

4.4 Conversion to binary

Sometimes one wishes to proceed in the reverse direction and, given a decimal number as a character string, convert to binary. Assume that the given number is an unsigned magnitude which does not exceed 65 535 so that it can be represented as a binary word. Let 34 567 be a possible example.

The first task is to change the ASCII code for the digit 3 into a binary number with the value 3. Then it must be multiplied by 10 000 since the contribution of the digit 3 to the decimal is 30 000. Similarly, the ASCII code for the digit 4 must be switched to the binary number for 4000. The addition of the two values leads to the binary number of value 34 000. Proceeding with each digit in this way we shall arrive at the desired quantity.

In setting up the module we shall assume that the character string starts at STRING and that the number of characters is in SLEN. The output is to be a word designated by VALUE. Since multiplication by an immediate value is not available there will have to be a declaration of the multiplier at some convenient stage, say

```
ten     db     10
```

The module could then look like

```
  conv_bin:   mov    value,0          ; start VALUE off
;                                       at zero
              xor    cx,cx            ; clear CX
              mov    cl,slen          ; CX is the number
;                                       of characters in
;                                       the string
              xor    si,si            ; clear SI
```

```
asccon:     xor    ax,ax           ; clear AX
            mov    al,string[si]   ; move ASCII digit
;                                    to AL
            sub    al,48           ; change ASCII code
;                                    to binary number
            cmp    cx,2            ; is it the last digit?
            jb     no_mult         ; if so, no multiplication
;                                    by 10 required
            push   cx              ; if not, save CX
            dec    cx              ; and reduce CX by 1
mult:       mul    ten             ; multiply digit by 10
            loop   mult            ; as many times as set by CX
            pop    cx              ; recover original value
;                                    of CX
no_mult:    add    value,ax        ; add contribution of
;                                    digit to VALUE
            inc    si              ; adjust SI for next
;                                    digit
            loop   asccon          ; go round again if
;                                    necessary
            ret                    ; return to caller
```

There is a new instruction here LOOP. This reduces CX by 1 and then checks to see if this new value of CX is zero. If it is not zero the program jumps to the address given in the LOOP instruction. If it is zero no jump is made.

The LOOP instruction is a member of a hierarchy of such instructions. The others also test the Zero Flag. The LOOPE (or LOOPZ) instruction decreases CX by 1. If the new CX is not zero and ZF = 1 jump otherwise make no jump. Similarly LOOPNE (or LOOPNZ) first diminishes CX by 1; then, if CX is not zero and ZF = 0, there is a jump to the given address.

To conserve registers with the module the requisite PUSHs and POPs would have to be added.

It is of interest to trace how this module handles 34 567 for which SLEN = 5. On reaching ASCCON, SI = 0, CX = 5, and VALUE = 0. Then the first digit 3, or rather the ASCII code for it (51), is transferred to AX. After subtraction of 48 AX contains the binary representation for 3. CX is compared with 2 and, since 5 is not below 2, control goes to the PUSH instruction which save the value 5 on the stack. Now CX is reduced to 4 and AX is multiplied by 10 giving 30. The LOOP knocks CX down to 3 and, since it is not zero, goes back to multiply by 10 again, so that AX = 300. The LOOP adjusts CX to 2 and makes another multiplication, so that AX = 3000. After another LOOP, CX = 1 and AX = 30 000. The next LOOP makes CX = 0, so that there is no jump. Instead, 5, the value of CX at ASCCON, is recovered from the stack and placed in CX. The next instructions result in VALUE = 30 000 (in binary), SI = 1, CX = 4, and a return to ASCCON.

This brings in the digit 4 and, because CX is one less, the procedure ends with VALUE = 34 000, SI = 2, and CX = 3 before going back to ASCCON. In this way, the binary representation is built up in VALUE. On the last pass SI = 4, CX = 1 and so there is a jump over the multiplication routine and the final digit 7 is added directly to VALUE. The subsequent LOOP changes CX to 0 and there is no jump back to ASCCON.

4.5 Stack linking

The use of the stack for saving registers has already been demonstrated. A more recondite involvement of the stack is to employ it to carry the parameters for a module as well as providing space for local variables. The stack pointer is obviously an important part of such a manoeuvre but another register, the base pointer, will be needed to keep track of the movements of the stack pointer i.e. a *link register* must be maintained.

The calling program is responsible for placing the parameters on the stack in the correct order before transferring control to the module. Suppose that there are two parameters, say ALPHA and BETA, to be passed. Whether their actual values or addresses pointing to them are placed on the stack will depend upon circumstances, but will not affect the principles in passing them so long as the module knows the rules. Suppose also that the module employs four words for local variables.

Consder the set of instructions after a CALL

```
push    bp          ; save BP on stack
mov     bp,sp       ; put stack pointer
                    ; in link register
sub     sp,8        ; 8 bytes on stack
                    ; for local variables
```

To see the significance of these instructions we examine the stack structure, remembering that *the stack grows towards lower addresses.* In Fig. 4.1 the stack is shown growing from address 2000H to 19E8H.

At first the content of SP is 2000H. The calling program inserts the two parameters, and SP moves to 19FCH, then issues CALL. This transfers the content of the instruction pointer IP to the top of the stack (SP becomes 19FAH) so that the program knows where to go on return from the module. The first of our three instructions saves BP on the stack and SP goes to 19F8H. Then SP is copied to BP so that BP contains 19F8H. The final instruction changes SP to 19F0H.

Thus the link in BP is a record of where the previous BP is and the adjacent address supplies the instruction pointer. In reality, we shall not know the stack addresses in advance but that does not matter with indirect addressing. One

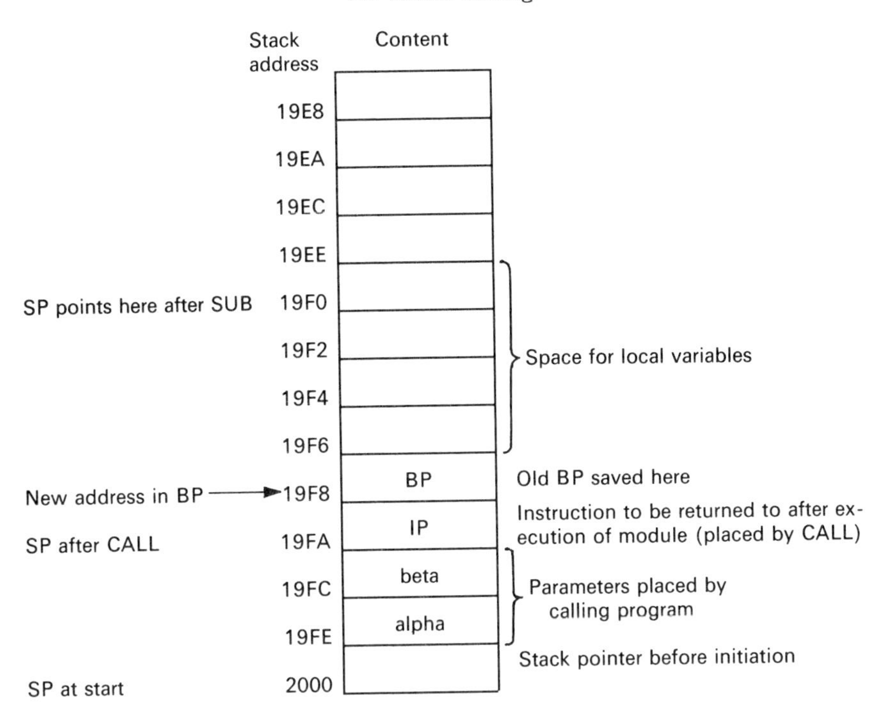

Fig. 4.1 Effects on stack

of the local variables can be referred to by [BP] − 4 and a parameter by [BP] + 6.

Now implement the instructions

```
        mov     sp,bp       ; make stack pointer
                            ; give link in BP
        pop     bp          ; restore original value of BP
```

The first instruction makes SP point to the address 19F8H so that this is now the top of the stack. The POP recovers the value of BP at the beginning and transfers SP to 19FAH. So an issue of RET will unstack the appropriate instruction pointer and execution of the program will resume at the correct stage on return from CALL. However, the stack pointer will be at 19FCH instead of 2000H leading to potentially uncontrolled growth of the stack if many calls are implemented. To avoid the possibility of such a disaster the instruction

```
add sp,4
```

should follow the CALL to the module. Alternatively, the ADD SP,4 can be omitted from the calling program and instead RET in the subroutine replaced by RET 4 since RET $2n$ moves the stack pointer back an additional n words on return.

It is clear that the technique of providing a link in BP back to an earlier step via the stack is very powerful. It allows a module to call another module, the parameters and local variables for the new module being located further up the stack, and the second module can call a third and so on. Thus, *nested calls* can be achieved to any level, subject only to the maximum permitted size of the stack. Since the module may call itself *recursive calls* may also be undertaken, though you must always take the precaution of ensuring that the recursion possesses a definite exit and does not end up in a permanent cycle.

4.6 Some general principles for links

The discussion in Section 4.5 of the effects of the two sets of instructions on the stack of Fig. 4.1 affords some general principles to be observed for links and stacks. Parameters intended for the module are passed via the stack, as are values going from the module to the calling program. The parameters may be the values of variables or their addresses.

Local variables are also placed on the stack. A local variable requires temporary storage while other processes occur. The storage is done on the stack primarily to evade any difficulty caused by interrupts. Of course, a module may have no local variables, in which case no space needs to be allocated for them.

The steps in setting up a link register may be summarized as follows:

(i) The caller prepares the parameters on the stack in the requisite order for the module.
(ii) A CALL is issued so that IP is conserved.
(iii) The module opens with 3 instructions putting a link in BP and allocating any desired space for local variables. Parameters are reached by positive offsets from BP whereas negative offsets are relevant to local variables.
(iv) The module stores the results of its calculations in locations previously agreed with the caller.
(v) The module finishes with 2 instructions restoring the original value of BP and with the stack pointer aimed at the original IP.
(vi) On RET control goes back to the calling program which then has the responsibility of processing any parameters produced by the module and returning the stack pointer to the address it was pointing to before the parameters were pushed on to the stack in (i).

4.7 A recursive program

As an illustration of the facility offered by linking via the stack we shall consider an example of recursion. In practice, programmers prefer to avoid recursion when there is a choice. A recursive program tends to use up space because a complete set of locations must be allocated every time it invokes itself. Also the overheads entailed in entering and leaving the module are likely to cause more time to be consumed than in a non-recursive routine. Nevertheless, there are some calculations which are awkward to program without recursion and these are therefore natural candidates for the technique. One such will now be discussed.

A rectangle is divided into squares and each square is marked with 0 or 1 (see Fig. 4.2). A coin is placed on one of the squares marked 0. It can move one square at a time, either vertically or horizontally (but not diagonally), subject to the condition that it never sits on a square marked 1. Can the coin reach the boundary? This is a kind of random walk problem with constraints.

A systematic approach is to try to move up one square. If that fails try one to the right. If that fails try one down. If there is still no success try one to the left.

Suppose that the coin succeeds in moving up one. Then unless it has reached the boundary, it faces the same problem as before though with a new base. So we shall have to call on the same process again i.e. recursion will be invoked.

To construct a program we shall assume that the marks 0 and 1 are contained in bytes forming an array labelled BOARD. Starting from 0 in the bottom left-hand corner the address of the squares relative to BOARD run steadily up each column (See Fig. 4.3 for the relative addresses for Fig. 4.2). A square can be identified either by its number or by the square at the bottom of a column together with the vertical offset from the bottom. Thus 18,2 would give square 20 and 9,5 square 14. The initial square of the coin will be specified in this format by stating the values of OPENX and OPENY so that the coin starts in the square numbered OPENX + OPENY.

1	1	1	1
1	0	1	1
1	0	0	0
1	0	0	1
1	1	0	1
1	0	1	1
1	0	0	1
1	0	0	1
1	1	1	1

FIG. 4.2 A board problem

It will be assumed that 0 and 1 have already been inserted in the array BOARD and that OPENX, OPENY are given. The program will be set out for Figs. 4.2, 4.3, but only slight changes are necessary to deal with any size of board. To call the module, which will be named SEEK, the following program is used

```
col_len   equ    9          ; specify the number of
                            ; squares per column
maxx      equ    27         ; specify number of right-
                            ; hand bottom square
exitm     db     0          ; declare marker for
                            ; success or failure
          mov    bx,openx   ; OPENX to BX
          push   bx         ; OPENX on stack
          mov    bx,openy   ; OPENY TO BX
          push   bx         ; OPENY on stack
          call   seek       ; look for route from
                            ; initial point
          add    sp,4       ; restore stack pointer
          cmp    exitm, 1   ; check for success
          je     printb     ; if so, jump
          . . .             ; otherwise follow instructions
                            ; to print 'NO EXIT'
printb:   . . .             ; instructions to print board
          end
```

8	17	26	35
7	16	25	34
6	15	24	33
5	14	23	32
4	13	22	31
3	12	21	30
2	11	20	29
1	10	19	28
0	9	18	27

Fig. 4.3 Addresses relative to BOARD

This program is fairly innocuous. It puts the opening parameters on the stack and then calls SEEK. On return it checks EXITM for 0 (failure) or 1 (success) and then prints the board. In the case of failure it prints the additional information 'NO EXIT'. Further instructions to confirm that the opening position is on a legitimate square and to print an error message otherwise would, no doubt, be an improvement.

More substantial effort has to go into SEEK. First, we must prevent the coin oscillating between two squares or going in a circle. Second, a record of any successful route or blind alley is desirable. Third, there will be two parameters and a local variable indicating the square occupied by the coin. Fourth, there are four sets of very similar instructions because of the four directions which may be open to the coin.

```
seek:   push  bp              ; save BP on the stack
        mov   bp,sp           ; establish link in BP
        sub   sp,2            ; two bytes for local variable
        mov   bx,[bp]+2       ; put Y in BX
        add   bx,[bp]+4       ; form X+Y
        mov   [bp]-2,bx       ; square number to local
                              ; variable
        mov   board[bx],2     ; indicating occupancy
                              ; by putting 2 in square
        cmp   [bp]+4,0        ; is coin in first column?
        je    found           ; if so, done
        cmp   [bp]+4,maxx     ; is coin in final column?
        je    found           ; if so, done
        cmp   [bp]+2,0        ; is coin in bottom row?
        je    found           ; if so, done
        cmp   [bp]+2,col_len  ; is coin in top row?
        je    found           ; if so, done
        jmp   up              ; if not on boundary try moving
                              ; coin
found:  mov   exitm,1         ; adjust marker for success
        jmp   got_exit        ; go to final instructions
up:     add   bx,1            ; try to move up
        cmp   board[bx],0     ; is there a block?
        ja    right           ; if so, go right
        mov   bx,[bp]+4       ; if not, put X
        push  bx              ; on stack
        mov   bx,[bp]+2       ; and then put
        add   bx,1            ; Y+1
        push  bx              ; on stack
        call  seek            ; call SEEK with new parameters
        add   sp,4            ; restore stack pointer
```

```
          cmp     exitm,1          ; check for success
          je      got_exit         ; if so, leave
right:    mov     bx,[bp]-2        ; otherwise recover square
                                   ; number for a move to right
          add     bx,col_len       ; move right
          cmp     board[bx],0      ; is there a block?
          ja      down             ; if so, go down
          mov     bx,[bp]+4        ; if not, put
          add     bx,col_len       ; new X
          push    bx               ; on stack
          mov     bx,[bp]+2        ; and put Y
          push    bx               ; on stack
          call    seek             ; call SEEK with new parameters
          add     sp,4             ; restore stack pointer
          cmp     exitm,1          ; check for success
          je      got_exit         ; if so, leave
down:     mov     bx,[bp]-2        ; otherwise recover square
                                   ; number for a move down
          sub     bx,1             ; move down
          cmp     board[bx],0      ; is there a block?
          ja      left             ; if so, try left
          mov     bx,[bp]+4        ; otherwise X
          push    bx               ; to the stack
          mov     bx,[bp]+2        ; and put
          sub     bx,1             ; Y-1
          push    bx               ; on stack
          call    seek             ; call SEEK with new parameters
          add     sp,4             ; restore stack pointer
          cmp     exitm,1          ; success?
          je      got_exit         ; yes, so leave
left:     mov     bx,[bp]-2        ; no, so try left
          sub     bx,col_len       ; adjust X
          cmp     board[bx],0      ; is there a block?
          ja      done             ; if so, no more tries
          mov     bx,[bp]+4        ; if not, put
          sub     bx,col_len       ; new X
          push    bx               ; on stack
          mov     bx,[bp]+2        ; and put Y
          push    bx               ; on stack
          call    seek             ; call SEEK with new parameters
          add     sp,4             ; restore stack pointer
          cmp     exitm,1          ; success?
          jne     done             ; if not, finished
got_exit: mov     bx,[bp]-2        ; recover square number
```

```
        mov   board[bx],3     ; mark successful route
done:   mov   sp,bp           ; return link to SP
        pop   bp              ; restore BP
        ret
```

The module starts by forming the square number and saving it in the space for the local variable. It then records its occupancy by changing the content from 0 to 2. After that there is a test for whether the square is on one of the four boundaries. If it is, the success marker is set and control transfers to the closing instruction GOT_EXIT. Here the square content is altered to 3 to indicate a successful outcome.

When the square is not on the boundary the coin tries to move. The four sets of movement instructions UP, RIGHT, DOWN, LEFT are similar in structure. First, the feasibility of a move is tested by seeing if the content of the destination is 0. If there is a barrier due to 1 or previous occupancy (2) the next movement is switched to, except after LEFT. If a move is permissible the new parameters are placed on the stack and a recursive call to SEEK is made. Success here leads to GOT_EXIT, and failure to the next movement.

On completion the board shows 0 on any unoccupied square, 2 on blind alleys and 3 on a successful route. Fig. 4.4 shows the result of the coin being initially at square 14 (see Fig. 4.3) of Fig. 4.2, revealing one blind alley and a successful route. On the other hand, if the starting square were 10, the program would report that no exit had been found.

From this example you can conclude that a recursive module will not terminate unless both

1	1	1	1
1	2	1	1
1	3	3	3
1	3	0	1
1	1	0	1
1	0	1	1
1	0	0	1
1	0	0	1
1	1	1	1

Fig. 4.4 Successful route

(i) there is at least one degenerate case in which the module can perform its designated task without invoking itself recursively,
(ii) it invokes itself in such a way that a degenerate case is attained in a finite number of steps.

In fact, our example contains two degenerate cases. One is when the coin occupies a boundary square so that an exit is immediately to hand. The second is that the module ceases to invoke itself recursively when every neighbouring square is blocked either because it is forbidden (marked 1) or because it has already been visited.

As regards (ii) the marking of a square to indicate occupation and not allowing a subsequent visit means that after a finite number of steps either all available squares have been exhausted or a boundary square is reached. In either event a degenerate case is entered. Consequently, our module is in accord with the principles (i) and (ii).

Exercises

In exercises 1–12 write assembly language for the stated operations

1. Subtract the word in memory at TOP from the word in AX.
2. If CX + DX is positive put it in AX, otherwise put it in memory at NEGATIVE.
3. The address of a signed number is in DI and the number occupies a byte. If it is positive do nothing but if it is negative change its sign and add 7AH. Leave any registers you use in their original state.
4. Redo Exercise 3, but when the byte is negative divide it by 7AH, returning the quotient to the location of the byte and the remainder in the adjacent byte.
5. An ASCII character is stored at CHAR. Transfer it to NEW and (a) if it is a control character clear the carry flag CF, (b) if it is a letter of the alphabet set CF and the direction flag DF and (c) for any other symbol raise CF and clear DF.
6. Repeat Exercise 5 with the overflow flag OF replacing CF.
7. An array of words consists of signed numbers. Its address is in SI and the number of words is in CX. Place the sum of the positive words in AX and the sum of the negative words in BX.
8. The address of an array of bytes is in SI and of another array of bytes is in DI. The length of the second array is in CX. Transfer the contents of the second array to the first array so that they occupy corresponding positions.
9. Convert a positive binary number in AL into its equivalent on the scale of 8.
10. Convert the word in AX to its decimal equivalent and store it in the array DECIMAL.
11. The unsigned magnitude in AX is a number of seconds. Convert it to hours and minutes, storing the result in HR and MIN, leaving the excess seconds in AX.

12. The date is given by the day of the week in AL (0 = Sunday, 1 = Monday, etc.), the day (1–31) in DL, the month (1 = January, 2 = February etc.) in DH, and the year (1980–2099) in CX. Store the date in the standard numerical form ./ ./. at DATE.
13. A module requires 3 word parameters from memory locations PARM1, PARM2, PARM3 as input. Place them on the stack before issuing CALL. The first few instructions of the module set up a link and provide for 3 local variables. List the instructions for this process and describe the effect on the stack, SP and BP of each one.
14. List the instructions when the module in Exercise 13 completes its task.
15. The expression $A^N = \text{expr}(A,N)$, where A and N are unsigned magnitudes, can be defined recursively by

$$\text{expr}(A,N) = A \text{ if } N = 1; \text{expr}(A,N) = A \times \text{expr}(A,N-1) \text{ if } N > 1.$$

Construct a module for calculating A^N when A is in AL and N in BL. Consider what precautions are desirable to keep the size within bounds.
16. The *Towers of Hanoi* is a game played with three poles and a set of disks, no two the same size, which fit on the poles. Initially, all the disks are on pole 1 (see Fig. 4.5) and they are to be moved to pole 2. Only one disk may be moved at a time and no disk may be above a smaller disk. The problem is thus one of moving N disks from a SOURCE pole to a DESTINATION pole, taking advantage of a SPARE pole. Observing that the bottom disk cannot be moved unless all the other disks are on the SPARE we try

(a) move N – 1 disk from SOURCE to SPARE, treating DESTINATION as a spare;
(b) move one disk from SOURCE to DESTINATION;
(c) move N – 1 disks from SPARE to DESTINATION treating SOURCE as a spare.

Write a recursive module to achieve this, counting the number of times a single disk is moved.

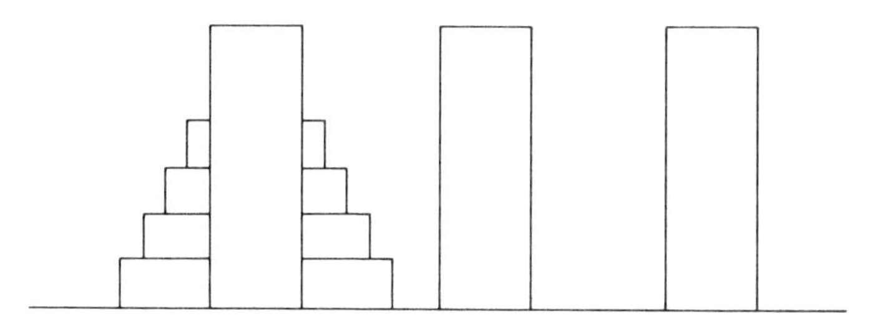

FIG. 4.5 The Towers of Hanoi

5

STRINGS

5.1 Introduction

Strings of characters are of common occurrence in computational work. Most high-level languages have special instructions for handling strings and we shall indicate how some of the standard ones can be emulated in assembly language. Sometimes it is convenient to manipulate the individual characters of a string and, on other occasions, it is more appropriate to treat the string as a whole.

Often a limit is set to the permitted length of a string, perhaps 255 characters, so a marker to signify the end of a string is desirable. Usually the marker is a carriage return (0DH in ASCII), very often followed by a line feed (0AH in ASCII). However, we shall mostly confine ourselves to examples in which the string is terminated by the carriage return alone.

The 8086 has a batch of instructions especially for coping with strings. Some particular conventions apply to these instructions. Firstly, the registers SI and DI are always used for addressing. The source is addressed by the SI register and the destination by the DI register. The address of the source is ascertained from SI and the Data Segment register (DS) for strings. In contrast, the destination address in string operations resides in DI and the Extra Segment register (ES). Unless there are good reasons to the contrary, it is sound practice to make DS and ES coincide.

Many of the string instructions have no operands. In the others the operands are invariably *dummies* whose sole purpose is to determine whether the operation is on bytes or words. There are two ways of setting up a dummy. The direct route is via PTR. Thus

```
stos        word ptr[di]
```

will transfer the word in AX to the destination whereas

```
stos        byte ptr[di]
```

would move the byte in AL. Alternatively, a dummy label can be introduced, as in

```
byte_dummy   rb     1              ; a dummy label for 1 byte
             stos   byte_dummy     ; store byte at
                                   ; address in DI.
```

In either case the assembler is informed that the byte form or the word form of

STOS is to be fed to the machine. The memory location for BYTE_DUMMY is not used at all and no characters are stored there.

After each string instruction there is updating of SI or DI or both depending on which was involved in the instruction. The adjustment alters the content of the register by 1 (for a byte operations) and by 2 (for a word). If the Direction Flag DF has been set to 1 the content of DI (and/or SI) is decremented by 1 (byte) or 2 (word). A register is incremented if DF = 0.

Although the source operand is normally addressed via SI and the DS register it is possible to designate a different segment by employing an *override prefix* e.g.

```
movs byte ptr[di],cs:byte ptr[si]
```

5.2 String storage

A string of ASCII characters is entering from port 127. The characters are destined for the memory location pointed to by DI. The maximum number of bytes for characters which can be stored is 0FFH. Process the characters until either the storage is full or a carriage return is received.

```
; Input: ASCII characters from port 127 to form
; string at [DI]. Exit: DI points to string with carriage
; return.
 input     equ    127            ; identify port of entry
;
 in_strg:  push   ax             ; save
           push   cx             ; registers
           push   di             ; used
           mov    ax,ds          ; ensure that
           mov    es,ax          ; ES=DS
;
           cld                   ; DF=0 so that string
                                 ; operations increment SI,DI
           mov    cx,0ffh        ; start counter with
                                 ; maximum number of characters
 cycle:    in     al,input       ; transfer character
                                 ; from port
           stos   byte ptr[di]   ; place byte at [DI] and
                                 ; increase DI by 1
           cmp    al,0dh         ; test to see if carriage
                                 ; return
           loopne cycle          ; if not decrease CX and, if
                                 ; CX≠0, jump
           cmp    cx,0           ; otherwise, test if CX=0
```

```
                jnz     over            ; if CX≠0, exit from cycle
                                        ; was after receipt of
                                        ; carriage return
                . . .                   ; otherwise, issue warning
                                        ; signal that no more
                                        ; characters could be received
                mov     al,0dh          ; and insert carriage
                                        ; return at
                stos    byte ptr[di]    ; end of string
;
     over:      pop     di              ; restore
                pop     cx              ; registers
                pop     ax
                ret                     ; go back to caller
```

The instruction CLD is a processor control instruction which clears the Direction Flag DF to zero. As a result, STOS automatically increases DI by 1 after it has stored the byte. If CLD is replaced by STD then DF is raised to 1 and STOS would decrease DI automatically.

The operand after STOS is a dummy, as already explained, and tells the assembler to arrange for a byte transfer. The dummy can be avoided by substituting the instruction STOSB. This has no operand but informs the assembler that a byte is involved. Similarly, STOSW can be used without an operand for a word transfer.

The program continually checks to see if the storage is full or the string has been terminated by a carriage return. If 0FFH characters are received without a carriage return, no further characters are accepted. A carriage return is added and provision has been made for the inclusion of a warning signal to make the operator aware that space has run out.

Finally, note that the POP instruction in OVER returns DI to its original position so that it is pointing to the opening location of the string.

Observe that by labelling the first instruction as IN_STRG a convenient way of calling the module from the main program is provided.

Once a string is in memory various facilities may be provided in high level languages for manipulating it e.g. determine its length, its leftmost or rightmost characters, or find the position of a substring. You may also wish to combine strings. The following sections indicate how some of these operations might be implemented in assembly language.

5.3 Length of string

We assume that the string has already been inserted in memory and is pointed to by DI. Its length is found by scanning the string until a carriage return is found. Remark that this will include all ASCII characters in the string so that any control codes (other than carriage return) will be counted as well as the characters which can be printed.

The length of the string will be calculated in CX and stored in SLEN. It is often convenient to keep the length in CX so this will be done in the following instructions. However, if CX must return to its original value then appropriate PUSH and POP instructions must be inserted and the length will be available only in SLEN.

```
; LEN_STRG. Entry: DI points to string. Exit: DI points
; to string; SLEN and CX give its length excluding
; carriage return but including other control codes
  len_strgd  dseg
  slen       rw     1
  len_strgc  cseg
  len_strg:  push   ax           ; preserve
             push   di           ; registers
             cld                 ; DF=0 so string operations
                                 ; increment SI,DI
             mov    al,0dh       ; put symbol for
                                 ; carriage return in AL
             xor    cx,cx        ; clear CX for counting
   cycle1:   scasb               ; take character in string
                                 ; from AL, then increase DI
             je     done         ; if they are same, finished
             inc    cx           ; otherwise increase count
             jmp    cycle1       ; and go round again
;
  done:      mov    slen,cx      ; store length
             pop    di           ; restore
             pop    ax           ; registers
             ret                 ; back to caller
```

The code and data segments have been given names to help with identification though this is not essential. The instruction SCASB scans the string in bytes and compares them with the byte in AL. After each comparison DI is increased since the Direction Flag DF has been cleared to zero. The label CYCLE1 has been used in case you wish to have IN_STRG and LEN_STRG in the same program. Each passage through the cycle increases the count by 1 to indicate that another character has been added to the list. Instructions to ensure ES=DS have been omitted.

Instead of SCASB we could have employed SCAS with a dummy e.g.

```
scas byte ptr[di]
```

When scanning a string in words, SCASW is available and compares a word in the string with a word in AX.

5.4 *Copy of a string*

To make a copy of a string we must treat it as a source. So its location must be pointed to by SI. A record of its length is also necessary and we assume that this has been placed in CX (perhaps by the module of the preceding section). The destination for the copy is pointed to by DI.

```
; COPY_STRG is entered with CX,SI set to string to
; be copied, DI its destination: on exit SI,DI are
; unaltered and DX contains length of string
  copy_strg:   push   cx         ; save registers
               push   si         ; used
               push   di
               push   ax
               push   ds
               push   es
               mov    ax,ds      ; ensure that
               mov    es,ax      ; ES=DS
               cld               ; arrange for autoincrement
               mov    dx,cx      ; put length of string
                                 ; in DX
               inc    cx         ; increase length so that CR
                                 ; is moved (if CR included
                                 ; in length omit this instruction)
  rep          movsb             ; move the string a byte
                                 ; at a time
               pop    es         ; restore
               pop    ds         ; registers
               pop    ax
               pop    di
               pop    si
               pop    cx
               ret
```

The program allows for the possibility that ES may not have been set to DS by the caller; if this is not so all the instructions involving AX, ES, and DS can be dropped.

MOVSB, a short way of saying Move String Byte, takes a byte from the source string, pointed to by SI, to the destination string, designated by DI, and then increases both SI and DI. MOVSW does the same for words, whereas MOVS works with dummies.

REP is one of five prefixes which can be attached to any string instruction. They cause the string instruction to be repeated as many times as the CX register indicates. The CX register is decreased by 1 on each iteration. Thus REP asks for a repetition of the string instruction whenever CX is not zero.

Similarly, REPE implies repetition if CX>0 and ZF=1 whereas REPNE gives repetition if CX>0 and ZF=0.

It follows that, although MOVSB copies only a byte at a time, the prefix REP enables the transfer of the whole string by means of one instruction.

5.5 *Left of string*

To extract a given number of characters from the left-hand end of a given string is a common requirement. For example the extraction of the 16 characters to the left of

MARY IS A BEAUTY AND IS A LADY

would yield

MARY IS A BEAUTY

as output.

SI points to the string and CX supplies its length. The destination is pointed to by DI and the number of characters desired is set in DX. In this case precautions against being asked to undertake the impossible are necessary. If the given string is shorter than the required length, then the entire string is returned. If the required length is 0 or negative then an error message will be issued by putting 0FFFFH in DX. Otherwise, the left-hand end of the string is returned and the requisite number of characters.

It will be assumed that ES and DS have been properly adjusted elsewhere. Consequently, when COPY_STRG is invoked from the preceding section it will be taken to be the version in which the instructions concerning AX, ES, DS are missing.

```
; LEFT_STRG is entered with CX,SI set to given string,
; DX the number of characters wanted and DI the destination: on
; exit DI points to the substring and DX contains the number of
; characters unless DX=0FFFFH when an error has
; been committed
left_strg:   cmp    dx,0         ; check if positive number
                                 ; of characters requested
             jg     can_copy     ; if positive go ahead
             mov    dx,0ffffh    ; otherwise put error signal
             jmp    finish       ; in DX and leave
;
can_copy:    push   cx           ; preserve
             push   si           ; registers
             push   di
             cmp    dx,cx        ; check if whole string wanted
             jge    lot          ; if so, copy
```

```
                mov     cx,dx           ; otherwise put desired
                                        ; length in CX
lot:            call    copy_strg       ; use copy module to
                                        ; transfer substring
                pop     di              ; restore
                pop     si              ; registers
                pop     cx
finish:         ret                     ; back to caller
```

You will notice that the main purposes of the instructions are to (a) confirm that a reasonable request has been made and (b) set up the registers so that COPY_STRG can be invoked. However, that leads us into a trap when CX excludes a carriage return. Then the instruction INC CX in COPY_STRG means that one more character has been returned than has been asked for. Moreover, if the substring is not the entire string, no arrangement has been made to terminate the substring with a carriage return should this be desired. It will be left as an exercise for you to think how to overcome these deficiencies.

5.6 *Right of string*

The aim of this section is to extract a given number of characters from the right of a string, e.g. the right 7 characters of

MARY IS A BEAUTY AND IS A LADY.

would make

A LADY.

the output. Again we have to beware that we are not asked to do the impossible.

```
; RIGHT_STRG is entered with CX,SI set to given string,
; DX the number of characters wanted and DI the
; destination: on exit DI points to the substring and
; DX contains the number of characters in it unless
; DX=0FFFFH when an impossible request has been made.
;
right_strg:     cmp     dx,0            ; check if positive number
                                        ; of characters asked for
                jg      go_on           ; if so, carry on
                mov     dx,0ffffh       ; otherwise put error signal
                jmp     exit            ; in DX and finish
;
go_on           push    cx              ; preserve
                push    si              ; registers
```

```
                push    di
                cmp     dx,cx           ; see if entire string wanted
                jge     whole           ; if so, copy
;
                add     si,cx           ; otherwise make SI
                                        ; point to carriage rcturn
                add     di,dx           ; and DI to right end
                                        ; of destination
                mov     cx,dx           ; length to be moved in CX
                inc     cx              ; allow for carriage return
                std                     ; set DF for autodecrement
    rep         movsb                   ; move substring, starting
                                        ; at the right
                jmp     tend:
;
    whole:      call    copy_strg       ; copy whole string
    tend:       pop     di              ; restore
                pop     si              ; registers
                pop     cx
    exit:       ret                     ; back to caller
```

The instruction STD allows us to move the substring by starting from the right instead of from the left as would be imposed by CLD. Do you think that this module suffers from any of the defects mentioned at the end of the preceding section?

5.7 Middle of string

The extraction of a portion of string which may not begin or end where the string does is the target of this section. The extracted substring is to be terminated by a carriage return. This time no subroutine will be called so that you can judge whether such a calling would be economical in this context.

As usual, SI will point to the source string and CX gives the number of characters in it excluding the final carriage return. DI is the pointer to the destination and DX contains the number of characters to be transferred. The position of the starting character of the substring is recorded in AX; AX=1 when the substring is to start at the first character of the source string. Since the substring cannot extend beyond the end of the source string, DX is set to the number of characters actually moved on completion of the transfer. Consistent with this DX=0 at the end if the requested transfer is impossible instead of indicating an error with an unexpectedly large number as in previous sections.

```
; MID_STRG is entered with CX,SI set to given string, DX
; the number of characters wanted, DI points to the
```

```
; destination and AX gives the starting character of the
; substring: on exit, the only change is that DX gives
; the number of characters transferred.
;
 mid_strg:  cmp   dx,0          ; see if positive number
                                ; of characters wanted
            jna   error           if not, leave
            cmp   ax,0          ; otherwise see if start of
                                ; substring is positive
            jna   error         ; if not, leave
            cmp   ax,cx         ; otherwise see if substring starts
                                ; beyond end of source
            ja    error         ; if so, leave
;
            push  cx            ; preserve
            push  si            ; registers
            push  di
            dec   ax            ; reduce AX by 1
            add   si,ax         ; SI points to first
                                ; character of substring
            sub   cx,ax         ; CX contains number of
                                ; characters available
                                ; for substring
            inc   ax            ; return AX to original value
            cmp   cx,dx         ; see if there are enough
                                ; characters for desired substring
            jb    adjust        ; if not, adjust DX
            mov   cx,dx         ; if so, put required
                                ; length in CX
adjust:     mov   dx,cx         ; DX shows number transferred
            cld                 ; arrange for autoincrement
rep         movsb               ; move the designated
                                ; number of characters
            mov   byte ptr[di],0dh
                                ; add the carriage return
;
            pop   di            ; restore
            pop   si            ; registers
            pop   cx
            jmp   last          ; leap over ERROR
;
error:      mov   dx,0          ; DX to show no characters
                                ; transferred
last:       ret                 ; back to caller
```

Only a small part of the program is occupied in moving the substring to its destination. The rest is concerned with checking that the operation is feasible and setting up the registers to accomplish the transfer. The values in AX,CX,DX have been treated as unsigned magnitudes (does this mean that the first four instructions are really superfluous?).

It will not have escaped your attention that the availability of MID_STRG renders the presence of LEFT_STRG and RIGHT_STRG unnecessary. For LEFT_STRG can be achieved from MID_STRG by putting AX=1 and RIGHT_STRG by making AX=CX−DX+1. However, that does mean committing an additional register and preparing it for entry to the module.

5.8 Combining strings

Frequently, one wishes to add new material at the end of a file. A simple version of this is to place one string immediately following another with no carriage return between them, a mechanism known as *concatenation.* Since two strings are involved extra registers must be brought into play and BP will point to the second string while BX supplies its length

```
; CAT_STRG is entered with CX,SI set to one string and
; BX,BP to the other: the pointer to the destination
; of the concatenated string is in DI. On exit the only
; change is that DX contains the length of the
; concatenation.
;
cat_strg:  push   ax          ; preserve
           push   cx          ; registers
           push   si
           push   di
;
           mov    ax,cx       ; save length of string 1
           call   copy_strg   ; copy string 1 to
                              ; destination
           mov    si,bp       ; SI points to string 2
           mov    cx,bx       ; CX is length of string 2
           add    di,ax       ; DI points to carriage
                              ; return of string 1 in
                              ; destination
           call   copy_strg   ; move string 2, its first
                              ; character obliterating
                              ; carriage return of string 1
           add    dx,ax       ; DX length of
                              ; concatenation
;
```

```
        pop     di          ; restore
        pop     si          ; registers
        pop     cx
        pop     ax
        ret
```

5.9 Search for a substring

A more recondite example concerning two strings is the analogue of the instruction INSTR in BASIC. The purpose of this instruction is to search a given string for the first occurrence of another given string. It is obvious that, if the second string is longer than the first, the search cannot succeed and so this eventuality must be eliminated first.

When the second string is the shorter the first character of the substring is compared with first character of the given string. If they do not agree the first character of the substring is compared with the second character of the given string. If there is no agreement the comparison is moved to the next character of the given string. This procedure is continued until either a match is found or the substring extends beyond the end of the given string and the search fails. Once a match has been discovered the second character of the substring is compared with its correspondent in the given string. If they are the same we proceed to the third character of substring and, if not, we go back to comparing the first character of the substring. In other words the substring is compared, character by character, with successive characters in the given string with the starting point of the substring being moved steadily along the string until either the search is called off or a match is secured.

If the search is successful the position of the first character of the given string where the substring occurs is to be returned in BX. In all other cases BX is to show 0.

```
; SEARCH is entered with CX,SI set to substring and
; DX,DI set to string to be searched. On exit BX contains
; the position within the searched string of the start of the
; first occurrence of the substring if found: otherwise
; BX shows 0.
;
  search:   xor     bx,bx       ; clear BX
            cmp     cx,dx       ; see if substring is longer
                                ; than given string
            ja      leave       ; if so, no more to do
;
            push    ax          ; save
            push    bp          ; registers
```

```
            push    cx
            push    dx
            push    di
            push    si
;
            mov     bx,cx       ; save length of substring
            mov     bp,di       ; save pointer to start
                                ; of string
            mov     cx,dx       ; length of string in CX
            sub     cx,bx       ; CX contains number of
                                ; characters in string extra to
                                ; substring
            inc     cx          ; CX=number of feasible
                                ; starting positions for substring
            cld                 ; arrange for autoincrement
            mov     al,[si]     ; put first character of
                                ; substring in AL
;
 cycle2:    repne   scasb       ; scan string for first
                                ; character of substring
            jcxz    fail        ; if repeat terminates because
                                ; starting positions exhausted
                                ; search fails
            push    cx          ; otherwise a match found
                                ; so save starting position
            push    di          ; reached and position of
                                ; next string character for
            push    si          ; comparison and pointer
                                ; to substring
;
            mov     cx,bx       ; restore length of substring
            dec     di          ; point DI to matched
                                ; character
 repe       cmpsb               ; compare substring character
                                ; by character with string
            pop     si          ; restore pointers in case
            pop     di          ; we need to try again
            jcxz    found       ; if repeat terminates because
                                ; CX=0 substring occurs
                                ; in string
            pop     cx          ; otherwise bring back
                                ; starting position
            jmp     cycle2      ; and try again
;
```

```
found:    pop   cx        ; keep stack in order
          mov   bx,di     ; BX points to character after
                          ; first match
          sub   bx,bp     ; BX gives position of first
                          ; occurrence of match
          jmp   close     ; leap over FAIL
;
fail:     xor   bx,bx     ; make BX zero
;
close:    pop   si        ; restore
          pop   di        ; registers
          pop   dx
          pop   cx
          pop   bp
          pop   ax
leave:    ret             ; back to caller
```

The instruction CMPSB, compare strings by bytes, checks the bytes pointed to by SI and DI, and adjusts the flags accordingly. It then updates SI and DI in accordance with the Direction Flag which, in our case, is arranged for autoincrement. The prefix REPE ensures that the comparison is repeated until two characters disagree or the counts run out which, in our case, happens when all members of the substring have found twins in the string being searched.

CMPSB could be replaced by CMPS, which compares bytes or words through dummies, by means of

```
cmps byte ptr[di],byte ptr[si]
```

Furthermore, CMPSW is analogous to CMPSB but compares words only.

Exercises

In exercises 8–12, carry out the operations described by an assembly program

1. Strings are always terminated by a carriage return followed by a line feed. What difference, if any, does this make to the programs LEN_STRG and COPY_STRG?
2. Modify LEFT_STRG so that the correct number of characters terminated by a carriage return is returned.
3. Modify MID_STRG to cover the case when AX,CX,DX are signed numbers.
4. Modify CAT_STRG so that there is a space between the end of the first string and the start of the second string, but no intervening carriage return.
5. Two strings are located in STRG1 and STRG2, and their lengths are given in the words at LEN1 and LEN2 respectively. Use your program of Exercise 4

to concatenate them with the result located at RSTRG and its length in RLEN.

6. Use the program of Exercise 5 to concatenate three strings STRG1, STRG2, STRG3.

7. Write a program to concatenate the first N characters of STRG1 followed by STRG2 in Exercise 5.

8. A string, referenced by DX and DI, contains a number of sentences. Extract the leftmost one and place it where SI points, recording its length in CX.

9. Count the number of spaces in a string, referenced by DX and DI, and place the result in CX.

10. Copy a string, referenced by CX and SI, into the location specified by DI, with the highest bit of each byte made zero so that every byte lies between 0 and 127.

11. Copy a string, referenced by CX and SI, into the location indicated by DI changing all lower case letters to upper case.

12. Repeat Exercise 11 but changing upper case letters to lower case.

13. A sequence of strings of ASCII characters is stored in the location HOME. Each string is terminated by a carriage return. The length of each string is given in consecutive bytes from SIZE and the number of strings is in the word NUM. Write a program which sorts the strings so that the lower ASCII representations appear before the upper. Temporary storage is available in 600 bytes at TEMP if desired.

14. Repeat Exercise 13 when each string is terminated by a carriage return followed by a line feed.

6
BINARY OPERATIONS

6.1 Logical and shift instructions

The 8086 possesses a number of instructions whose primary purpose is to operate on bits rather than the bytes and words that have concerned us hitherto. XOR is one example that has already been encountered because of its value in clearing a register. Wherever the bit pattern has to be examined (perhaps to check sign) or has to be altered, as when packing (see Section 6.3), one is likely to come across these instructions. Some simple illustrations will now be discussed.

Suppose that we wish to leave the highest bit of a byte unaltered but change all the other 1s and 0s to 0s and 1s respectively.

```
; SWITCH is entered with a byte in AL. On exit the byte in
; AL has bit 7 unchanged but all other 1s and 0s have become 0s and 1s.
;
switch:   xor    al,80h  ; change bit 7 from 1 to 0
                         ; or vice versa
          not    al      ; form ones-complement
          ret            ; back to caller
```

The binary 80H has a 1 in bit 7 and a 0 in bits 0–6. With XOR a 0 has no effect on the bit in the destination whereas a 1 causes a bit to be set if it is zero, otherwise it is cleared. Thus the effect of the XOR is to change only bit 7. The instruction NOT performs the logical complement i.e. it converts every 1 to 0 and every 0 to 1. Hence, the NOT instruction restores bit 7 to its original value and changes all the other bits as required.

To clear part of a byte, AND is the appropriate instruction because a 0 always clears whereas a 1 leaves a bit unaltered. Therefore to clear the second, fourth, . . . bits of a byte we should AND it with 10101010B or 0AAH. For example

```
and    al,0aah
```

would effect the desired clearance in AL.

The concatenation of bits from two words can be achieved via AND and OR. Suppose that we wish to concatenate bits 4–15 of AX with bits 0–3 of BX. Clear bits 0–3 of AX by means of AND. Then, since a 0 in OR always reproduces the other bit, an OR with BX will produce the desired concatenation provided that bits 4–15 of BX are cleared first. To clear bits 0–3 of AX,

AND with 1111 1111 1111 0000 or 0FFF0H. The fragment will have the form

```
        and     ax,0fff0h   ; clear bits 0–3 leaving
                            ; rest unaltered
        and     bx,0fh      ; clear bits 4–15
        or      ax,bx       ; concatenate, result
                            ; being in AX.
```

Moving bits around is a more complicated operation unless bytes or words are to be interchanged. Then XCHG is appropriate, e.g.

```
        xchg    ah,al
```

exchanges the contents of AH and AL. However, XCHG cannot be used if we want bits 0–2 of AL to move to bits 5–7 while bits 3–7 go to bits 0–4. For this rotation is suitable; for instance

```
        ror     al,1        ; rotate bits in AL to
        ror     al,1        ; the right in a circle
        ror     al,1        ; three times
```

would achieve the desired interchange.

If we had wanted to move bits 0–4 to bits 3–7 it would be more economical to deploy ROL, rotate to the left, three times rather than ROR five times. Multiple rotations can be encompassed in an instruction as follows

```
        mov     cl,3        ; put rotation count in CL
        ror     al,cl       ; do 3 rotations to the right
```

For a small number of rotations this is a less efficient technique than repeated instructions to rotate one position.

6.2 Changing flags

Direct access to the flag registers is restricted. Instructions for setting or raising and clearing the direction flag DF have already been encountered. There are similar instructions STI, CLI for the Interrupt Enable Flag IF. Three instructions affect the Carry Flag CF directly, namely, CLC which clears the flag (CF = 0), STC which raises the flag (CF = 1) and CMC which takes the ones-complement of CF i.e. changes it from 1 to 0 or vice versa.

Four other flags can be adjusted in an indirect fashion. They are the Sign Flag, SF, the Zero Flag, ZF, the Auxiliary Carry Flag, AF, and the Parity Flag, PF. The Carry Flag can also be reached by this route. Suppose that the ones-complement of SF, ZF, AF, PF is required while CF is to be unaffected. Here is how it might be done:

```
        lahf                ; load AH from flags
        xor     ah,0d4h     ; take ones-complement
```

```
                        ; of S,Z,A,P
        sahf            ; transfer AH to the flags
```

The instruction LAHF moves a copy of bits 0–7 of the flags register (see Fig. 2.3) to AH so that it contains SZ?A?P?C. The number of 0D4H in binary is 11010100 so that the XOR finds the ones-complement of SF,ZF,AF, and PF while leaving the other bits alone. Then SAHF puts the contents of AH back in the flags under the same convention as to position as LAHF.

Notice that the state of all the flags can be preserved on the stack by PUSHF and recovered from the stack by POPF.

6.3 Packing

So far we have been considering each byte of information as occupying its own byte or storage without paying too much attention to what was being represented. This might be termed the natural or *unpacked* way of storing data. When processing bulky data it may be advantageous to try to store the items more economically. In some circumstances, it may be feasible to accomplish this by squeezing several pieces of data into each storage location.

Consider the storage of an array of ten single-digit positive integers. In the natural mode this would occupy ten bytes of storage, e.g.

4	0	1	6	1	3	3	2	1	7

However, a byte is certainly capable of representing a two-digit integer, so that two components of the array could be squeezed into each store location, enabling the space for the array to be reduced to five bytes, e.g.

40	16	13	32	17

In fact, a little more compression can be achieved by going to words, which can represent five-digit numbers (so long as they are less than 64 k), e.g.

40161	33217

which brings the storage requirement down to two words or four bytes.

For both cases the array is said to be *packed*. It is still possible to get at the individual components of a packed array, but the difficulty of access is increased because there has to be a conversion process to break the stored items down into their original constituent digits. In addition, there may be extra housekeeping, as in the word storage example above, to ensure that the capacity of the store is not exceeded. The conclusion is that the saving of storage space has been accompanied by a slowing down in access to the original digits and a longer program in machine-code.

Whether the trade-off between diminished storage requirements and increased machine effort is worthwhile can only be judged after a thorough examination of the context. The 8086 provides help with packing by offering special instructions for handling *binary-coded decimal.* They have the distinctive feature that the arithmetic handles two digits at a time instead of one. Accordingly, packed decimals can be coped with in half the number of instructions for manipulating the equivalent unpacked decimals. Despite the terminology, the decimals are integers. No decimal point is supplied. If you want one you must make your own arrangements, or turn to floating point as in Section 6.9.

In binary-coded decimal a decimal digit is represented, not by a byte, but by 4 bits. The standard form is used, set out there for convenience:

Decimal digit	4-bit code	Decimal digit	4-bit code
0	0000	5	0101
1	0001	6	0110
2	0010	7	0111
3	0011	8	1000
4	0100	9	1001

If four zeros are placed on the left of the 4-bit code then the standard byte representation of the decimal digit is obtained; this is known as *unpacked binary-coded decimal* or *unpacked BCD*, for short. On the other hand, if two 4-bit codes are placed side-by-side in a byte, there are two binary-coded decimals in the byte and this is called *packed BCD.*

6.4 Changing ASCII digits to packed BCD

The byte representing an ASCII decimal digit consists of the four bits 0011 followed by the 4-bit BCD. Therefore an ASCII digit can be converted to BCD by stripping off the leftmost 4 bits. One way of doing this is to perform AND with 0FH since the 0 clears the left four bits while the F leaves the right four bits. Another way, pertinent to packing, is to employ a left shift so that the BCD occupies the left four bits whereas the right four bits all become zero.

A six-digit decimal is given by six ASCII bytes in the location ASC. It is to be converted to packed BCD in three bytes at DML, with the digits running in the opposite order.

```
; ASC_BCD is entered with ASC containing six ASCII
; digits in bytes. On exit the three bytes at DML are filled with
```

```
; packed BCD with the digits in opposite order.
;
        extrn     asc:byte             ; the memory locations ASC,
        extrn     dml:byte             ; DML are declared elsewhere
                                       ; but are available here
        asc_bcd:  push   ax            ; preserve
                  push   cx            ; registers
                  push   di
                  push   si
;
                  lea    di,dml+2      ; put the address of
                                       ; destination in DI, ready to
                                       ; arrange digits in opposite order
                  lea    si,asc        ; address of source in SI
                  mov    cx,3          ; prepare counter
        cyc:      jcxz   fin           ; finish when 3 bytes packed
                  mov    al,[si]       ; insert digit
                  and    al,0fh        ; mask upper 4 bits
                  mov    ah,[si]+1     ; get next digit
                  shl    ah,1          ; shift 4 bits on
                  shl    ah,1          ; right to 4 bits on left
                  shl    ah,1          ; and fill right 4 bits
                  shl    ah,1          ; with zeros
                  or     ah,al         ; pack the 2 digits
;
                  mov    [di],ah       ; store result
                  dec    di            ; decrease DI
                  add    si,2          ; adjust source pointer
                  dec    cx            ; decrease counter
                  jmp    cyc           ; go round again
;
        fin:      pop    si            ; restore
                  pop    di            ; registers
                  pop    cx
                  pop    ax
                  ret                  ; back to caller
```

EXTRN is an assembler directive. It tells the assembler that ASC and DML can be used in this subroutine but they are defined in some other program. The term BYTE after the colon informs the assembler that the memory locations were declared to be in bytes when they were defined. The types WORD and DWORD should be used instead when appropriate. An EXTRN directive for a given symbol must appear within the segment in which the symbol is defined in some other module. It is deployed here to show a self-contained subroutine can be written.

Advantage has been taken of both methods of stripping off the unwanted bits. For the first digit (in AL) the left 4 bits are masked i.e. cleared by means of AND. In the second digit they are shoved out by 4 shifts to the left via SHL. An OR then packs the two digits into a byte. Note that the four shift instructions could be replaced by SHL AL,CL with CL = 4, but since CX is already acting as a counter that course is not open.

Putting the packed digits in the opposite order to the original was merely intended as an exercise. The Intel convention on storage of packed BCD is to take the digits in pairs and then store them in the opposite order. Thus 123456 would be stored as 56 34 12.

6.5 Addition of packed BCD

Addition of packed BCD is accomplished by relatively simple instructions but understanding what is going on is less straightforward. The reason for the complication is that the addition is carried out as a *binary* operation and then the result is corrected to packed BCD form through the instruction DAA (Decimal After Addition). Consider the three examples

(a)	31	0011 0001	(b)	35	0011 0101	(c)	49	0100 1001
	48	0100 1000		58	0101 1000		58	0101 1000
		0111 1001			1000 1101			1010 0001

where the result of the binary additions of the packed BCD representations of the decimal numbers on the left is shown. In (a) the right four bits give 9 and the left four bits 7; it correctly offers 79 as the packed BCD answer. In (b) the right four bits stand for 13 which is not a single digit. To make it a single digit subtract 10 and carry 1 to the digit in the left four bits. An equivalent operation is to add 6 in binary since this does the carry automatically and this is what DAA does so that

```
1000 1101
     0110
---------
1001 0011
```

Now the right four bits give 3 and the left four 9 leading to the correct answer 93.

At first sight there appears to be no trouble with the right four bits of (c) since they supply the single digit 1 until it is realized that the digit ought to be 7. The cause of the difficulty is that 9 + 8 cannot be represented by four bits. It involves a carry to the left four bits leaving behind 1. To indicate that this has happened the processor raises the Auxiliary Carry Flag AF to 1. This tells DAA to add 6 to the right four bits with the result 1010 0111. However, the left four bits do not provide a single digit but 10. Again DAA adds 6 to the left four bits so that

```
   1010 0111
   0110
  ----------
1 |0000 0111
```

Thus the byte shows 07 while the 1 which is carried is set in the carry flag CF.

Consequently, what DAA does is to examine the right four bits. If they exceed 9 or AF = 1 it adds 6 to them. Then it looks at the left four bits; if they exceed 9 or CF = 1 it adds 6 to them. The instruction DAS performs a corresponding adjustment for subtraction.

The mechanism adopted by DAA explains the Intel convention for storage of packed BCD. By working from the lower end to the upper end the flags AF and CF are always available to ensure that the correct carry goes forward.

As an illustration we consider the addition of two 4 digit numbers stored in packed BCD format at ALPHA and BETA respectively. The answer is to be placed in RESULT. Storage is in accordance with the Intel convention so that if 1234 is represented in ALPHA,34 will be in the byte at ALPHA and 12 in the byte at ALPHA + 1. We give only the skeleton of the program, without any preliminaries or closure.

```
      lea   si,alpha       ; address of ALPHA to SI
      lea   di,beta        ; address of BETA to DI
;
      mov   al,[si]        ; first pair of digits to AL
      add   al,[di]        ; binary addition
      daa                  ; adjust AL to packed BCD
      mov   result,al      ; store low pair of digits
                           ; of answer
;
      mov   al,[si] + 1    ; next pair of digits
      adc   al,[di] + 1    ; add with carry in case
                           ; there is carry from first
                           ; addition
      daa                  ; adjust AL to packed BCD
      mov   result + 1,al  ; store answer
```

You may think it wise to have a final check of CF to see if the sum could not be accommodated in four digits.

6.6 Unpacked BCD

The 8086 does not possess instructions for the multiplication or division of packed BCD so attention is turned to unpacked BCD in which bits 4–7 are always zeros. Again, arithmetical operations are performed in binary accompanied by suitable instructions for correction. Unpacked addition has similar problems with the right four bits as packed. So the instruction AAA (Adjust After Addition) is provided for amendment. It operates in a similar fashion to DAA with the right four bits, namely, if these bits exceed 9 or the Auxiliary Carry Flag is set it adds 6. A new feature is that a 1 may now appear in bits 4–7 so that the unpacked format is lost. When this occurs AAA clears AH, then adds 1 to AH, clears bits 4–7 of AL, and sets CF = 1. Remark that when the four right bits lie in the range 0–9 and AF = 0 then CF = 0. AAS undertakes a similar function for subtraction.

As an example consider the addition of a 6-digit number in unpacked BCD in six bytes at ALPHA to a similar 6-digit number at BETA. Recall that, according to convention, the number 123456 will be stored as 6 5 4 3 2 1 . The result of the addition is to be placed in 7 bytes (allowing for a possible carry) of unpacked BCD at SUM. In the following skeleton there is no explicit mention of AH but it must be remembered that it can be adjusted by AAA so it would not be safe to assume that this register was unaffected by the subroutine.

```
        lea   si,alpha              ; address of ALPHA to SI
        lea   di,beta               ; address of BETA to DI
        xor   bx,bx                 ; clear BX ready to act as
                                    ; pointer for SUM
        mov   cx,6                  ; set up counter
;
        clc                         ; clear CF ready for first
                                    ; add with carry
round:  mov   al,[si]               ; insert digit of ALPHA
        adc   al,[di]               ; add with carry digit of
                                    ; BETA
        aaa                         ; adjust to unpacked BCD
        mov   sum[bx],al            ; store digit
        inc   si                    ; prepare
        inc   di                    ; for next
        inc   bx                    ; digit
        loop  round                 ; decrease CX and if not zero
                                    ; repeat addition
;
        mov   byte ptr sum[bx],0    ; put 0 in highest byte
                                    ; of SUM
        jnc   exit                  ; if no carry, finished
        mov   byte ptr sum[bx],1    ; otherwise put 1 at top of SUM
exit:   ret
```

6.7 *Multiplication*

Multiplication of unpacked BCD is limited to the product of two digits. For the product of a number with more than one digit with a number containing a single digit we have to multiply each digit of the first number by the single digit separately and keep adding the results. If the second number had another digit the exercise would have to be repeated for the extra digit. Ensuring that each contributor is added with the right power of 10 can therefore be quite tiresome.

Furthermore, since the multiplication is in binary, correction to unpacked BCD after it is necessary. This is accomplished by AAM (adjust after multiplication). What this instruction does is to divide AL by 10 and place the quotient in AH whereas the remainder goes in AL.

The simplest illustration, displaying some of the housekeeping, is to multiply the 2-digit number in ALPHA by a single digit in BETA with the answer going to PRODUCT. Three digits must be allowed for at PRODUCT so three bytes must be declared there.

```
lea    si,alpha          ; address of ALPHA in SI
mov    bl,beta           ; multiplier in BL
mov    al,[si]           ; first digit of ALPHA in AL
mul    bl                ; multiply digits as unsigned
                         ; magnitudes, result will be
                         ; confined to AL
aam                      ; adjust to unpacked BCD
mov    product,al        ; store lowest digit
mov    product + 1,ah    ; store temporarily next digit
mov    al,[si] + 1       ; get next digit of ALPHA
mul    bl                ; multiply by BETA
aam                      ; adjust to unpacked BCD
mov    bh,ah             ; keep a copy of AH
add    al,product + 1    ; add in part of product already
                         ; saved
aaa                      ; adjust to unpacked BCD after
                         ; addition
adc    bh,0              ; add any carry to third digit
mov    product + 1,al    ; store
mov    product + 2,bh    ; answer
```

6.8 *Division*

Since division can be regarded as a kind of inverse process to multiplication you might expect that any correction instruction has to be undertaken before the binary division and this is indeed the case. The pertinent instruction is

AAD (adjust for division)—it might have been helpful if this had been named ABD as a constant reminder that it occurs before division. AAD works on AX, on the basis that AH and AL each contain a digit in unpacked BCD format with the digit in AH one order above AL. It multiplies AH by 10 and adds it to AL thereby creating in AL a binary representation of the 2-digit number. AH is then cleared. In other words AAD is the reverse of AAM.

Let us divide ALPHA by BETA, as defined in the preceding section, putting the quotient in QUOT and the remainder in REM.

```
mov   al,alpha        ; low-order digit in AL
mov   ah,alpha + 1    ; high-order digit in AH
mov   bl,beta         ; divisor in BL
aad                   ; prepare for division
div   bl              ; divide, getting quotient in AL
                      ; and remainder in AH
mov   rem,ah          ; store remainder
xor   ah,ah           ; clear AH
aam                   ; take binary in AL to two
                      ; unpacked BCD
mov   quot,al         ; store
mov   quot + 1,ah     ; quotient
```

After dealing with the remainder we have to change the binary quotient into two digits of unpacked BCD. This can be achieved by AAM provided that AX has the structure that occurs after a multiplication, which is easily arranged by clearing AH.

6.9 Binary floating point

Up to now only integers have been the subject of study. There are, naturally, many circumstances when the numbers and measurements we handle are not exact integers i.e. we must be prepared to deal with *real numbers.* Not all real numbers can be stored in a computer because there is a limit on the number of digits. Consequently, there is a restriction on the *magnitude* of a real number that can be stored as well as restraints on the *precision* with which a real number can be represented. The limitation on magnitude means that the danger of overflow is ever-present while the defect in precision may render the results of calculation practically useless.

To exploit the available precision to its full, only most significant digits are stored. The position of the decimal point relative to the stored digits is recorded separately as a *scale factor* or *exponent.* A positive exponent moves the point to the right whereas a negative exponent moves the decimal point to the left. For example, in a computer capable of storing a real number to 4 decimal digits:

0.0088	is stored as 8.800 with an exponent of -3
8.8	is stored as 8.800 with an exponent of 0
8800	is stored as 8.800 with an exponent of 3
9.75321	is stored as 9.753 with an exponent of 0
975382	is stored as 9.754 with an exponent of 5.

This representation is known as *floating point* because the number of digits stored is fixed and the decimal point floats relative to the stored digits. Therefore the position of the decimal point has to be kept sight of and taken into account in calculations.

A real number with more significant digits than can be stored loses precision as is shown by the last two examples above. The cutting down can be effected by *rounding*, as above, or by *truncation*, in which the extra digits are just discarded. Evidently, rounding is responsible for less inaccuracy in the representation than truncation. Since computers actually store in binary notation all real numbers which do not have an exact binary representation (which includes all irrational numbers) or possess too many significant figures are subject to *representational error* in the computer. This immediately forces error in any computations involving them, with the consequence that the results may be even less accurate than the operands i.e. *computational error* is introduced.

You are perhaps familiar with some of these points from PASCAL, or other high-level language, in which the notation for the above five examples is 8.8E-3, 8.8E0, 8.8E3, 9.753E0, 9.754E5. The number after the E tells you how many powers of 10 you multiply the first number by.

The same concept can be applied to binary real numbers. Thus .0011 would be stored as 1.1 with exponent -3, but this time the exponent tells you how many powers of 2 to multiply by; equivalently, it specifies the number of places the binary point is to be shifted. In floating point format the stored bits always start with a 1 so a storage space can be saved by recording only the bits after the binary point. However, a bit must always be devoted to the sign. The exponent may also have either sign but, once its range has been decided, it is given an offset so that it is always positive.

In 32-bit storage a standard format is to use bits 0–22 for the stored bits (after the binary point), bits 23–30 for the offset exponent and bit 31 for the sign. The offset exponent has 8 bits and so can range from 0 to 255. However 0 and 255 are reserved; 0 to indicate a number that is identically zero and 255 to cope with exponent overflow.

The offset is fixed at 127 so, since the offset exponent goes from 1 to 254, the actual exponent ranges from -126 to 127. Thus the largest number which can be stored in the above format is a little under 2^{128} or about 3.4×10^{38}. Similarly, the smallest non-zero number is 2^{-126} or about 1.18×10^{-38}.

The exponent is always an integer and therefore always dealt with exactly, except in cases of overflow. Hence any errors stem from the stored bits. In other words, the precision is that of 24 bits (including the omitted 1) which corresponds to 7 decimal digits.

Whatever the virtues of the standard format for storage purposes it is not terribly convenient for manipulation, partly because we have to keep supplying the missing 1 but mainly because we would like to have some extra bits available in an effort to ensure that the computational error is not appreciably worse than the representational error. It is therefore not uncommon to set aside two sections of memory each longer than 32 bits for carrying out arithmetic operations in floating point. These might be called the *Floating Point Register* and *Floating Point Accumulator* respectively.

In order to keep the complication down without obscuring the principles we shall make the artificial assumption that the number of stored bits after the binary point is 11. Incorporating the compulsory leading bit of 1 we have 12 bits for precision. The offset exponent will occupy 8 bits and follow the rules enunciated earlier. We now create a Floating Point Register from two consecutive words. In word 0 the structure will be

	0 1 b b b b b b		b b b b b a a a	
Bit number	15	8	7	0

where b stands for a stored bit, the 1 in position 14 is the compulsory leading 1 and the a is an extra bit to help with accuracy. When a number is transferred from store to this Register each a is put equal to zero. In the second word the left-hand byte is either 1 or 0 and takes care of the sign (0 signifying positive) whereas bits 0–7 hold the offset exponent.

The Floating Point Accumulator is constructed in the same way. When a number is transferred from it to store the question arises of how to use aaa for rounding. Obviously if aaa is greater than 100 we add 1 to the b in position 3 and if aaa is less than 100 we do nothing. If aaa is 100, follow the 'even' convention i.e. if the b in position 3 is 1 add 1, otherwise do nothing.

It has already been explained that the values 0 and 255 of the offset exponent are kept for special purposes. To be more precise an offset exponent of 255 and an all-zero word 0 in the Floating Point Accumulator will convey the information that an overflow has occurred in a floating point operation. Likewise an offset exponent of 0 and an all-zero word 0 records either that zero was the outcome or that an underflow occurred in a floating point operation.

There is a mathematical co-processor, the 8087, associated with the 8086 which handles floating point efficiently but its subtleties are outside the scope of this book. We shall now indicate the tricky footwork called into play in keeping track of things like a Floating Point Register where, in operations, different parts have to be treated in totally different ways.

6.10 Multiplication in floating point

To multiply numbers in floating point format add the exponents and form the product of the other two factors. There are two points to notice about the addition of the exponents. Firstly, their sum may be too large to lie in the permitted range. Secondly, when two offset exponents are added the sum contains the offset twice, once from each exponent, so one offset must be subtracted from the sum in order that it correctly represents an offset exponent.

With regard to the other two factors each is a number not less than 1 so that the product can exceed 2. When that happens an adjustment to the exponent is required. In fact, if the binary point between bits 13 and 14 is ignored, two numbers between 2^{14} and $2^{15}-1$ are multiplied. Their product is therefore bound to be at least 2^{28} and may surpass 2^{29} but cannot reach 2^{30}. It is the case when 2^{29} or greater transpires that must be recognized by an increase of 1 in the exponent. It is accounted for in the procedure for placing the leading 1 in bit 30 in order to conform to our format that there is one 0 ahead of the leading 1.

Much of the following program is occupied with attending to these matters while relatively few instructions are devoted to the actual multiplication. The purpose of the program is to multiply the number in the Floating Point Register by that in the Floating Point Accumulator, leaving the result in the Floating Point Accumulator. The addresses of the Register and Accumulator are contained in SI and DI respectively. No attempt has been made to preserve the registers which are employed in the multiplication.

```
flpt_mul    mov   cx,[si]+2     ; get exponent (with
                                ; sign) from Register
            mov   dx,[di]+2     ; and Accumulator
            cmp   cl,0          ; see if either exponent
            jz    ans_0         ; zero: if so product
            cmp   dl,0          ; is zero: adjust
            jz    ans_0         ; Accumulator and leave
            xor   ch,dh         ; obtain the sign of product
            mov   ah,ch         ; store sign temporarily
;
            xor   ch,ch         ; clear CH and DH ready for
            xor   dh,dh         ; any carry from addition
            add   cx,dx         ; of exponents
            sub   cx,7fh        ; remove extra offset
            cmp   cx,0feh       ; check for overflow
            jbe   too_small     ; if none go to check
                                ; for smallness
```

```
too_large:  mov   word ptr[di],0  ; otherwise clear bottom
                                  ; of Accumulator
            mov   word ptr[di]+2,0ffh
                                  ; and set exponent
                                  ; to 255 to signal overflow
            ret                   ; back to caller
;
too_small:  cmp   cx,0            ; see if exponent is
            ja    next            ; positive: if so go to next stage
                                  ; otherwise underflow or 0 so
                                  ; adjust Accumulator
ans_0:      mov   word ptr[di],0  ; clear Accumulator
            mov   word ptr[di]+2,0
            ret                   ; and back to caller
;
next:       mov   ch,ah           ; recover sign
            mov   ax,[si]         ; prepare for
            mov   bx,[di]         ; multiplication
            mul   bx              ; result in DX,AX
            shl   dx,1            ; start moving leading 1
            rol   ax,1            ; prepare bits in AX for
                                  ; transfer to DX
            test  dx,4000h        ; see if 1 in bit 14
            jz    shift           ; if not, shift again
            cmp   cl,0fdh         ; if so, check if exponent
                                  ; can cope with increase
            ja    too_large       ; if too big: overflow
            inc   cl              ; otherwise, increase exponent
            and   ax,1            ; clear unwanted bits
            jmp   trans           ; and transfer remainder
;
shift:      shl   dx,1            ; another shift should
            rol   ax,1            ; be enough
            and   ax,3            ; keep last 2 bits only
trans:      or    dx,ax           ; transfer bits from AX
                                  ; to tail of DX
            mov   [di],dx         ; store product
            mov   [di]+2,cx       ; store exponent and sign
            ret                   ; finish
```

The instruction TEST performs an AND on the two operands but records the result only in the flags, and does not alter the operands.

It could be argued that the accuracy has been impaired because rounding was ignored in favour of truncating the product to the right size for passing to the Floating Point Accumulator. However, the number transferred to the Accumulator does possess 3 accuracy bits so that the error is not as serious as might be first imagined.

6.11 Addition in floating point

Adding two numbers in floating point format is more complicated than multiplication because it must first be arranged that both have the same exponent. One or other of the exponents will in general have to be changed. We choose to keep the larger exponent unaltered in the interests of accuracy. So the smaller exponent is augmented until it is equal to the other. At the same time the associated stored bits must be shifted to the right in compensation. Of course, if the shift is large enough all the stored bits will be lost. In that case, one of two numbers in the addition is effectively zero and the sum is just equal to the number with the larger exponent.

Even when the exponents have been taken care of it is not quite straightforward to carry out the addition. Dealing with the sign requires a little thought and the result may not have the leading 1 in the correct position; overflow is not a problem here because our representation starts with 0. However, in moving the leading 1 to its rightful location we have to modify the exponent with the possibility that overflow or underflow may occur. How these points are looked after will be clearer when you have had a chance to examine the program.

The two numbers to be added are in standard format in the Floating Point Register and Accumulator respectively, whose addresses are to be found in SI and DI. The result is to be put in the Accumulator. Again, conservation of registers will be neglected.

```
flpt_add:   mov   cx,[si]+2    ; get Register exponent
            cmp   cl,0         ; if exponent 0, sum
            jnz   more         ; is already in
            ret                ; the Accumulator
;
more:       mov   dx,[di]+2    ; get Accumulator exponent
            mov   ax,[si]      ; and rest of Register
            mov   dx,[di]      ; and Accumulator
            cmp   dl,0         ; if Accumulator zero
            jnz   cont         ; sum is Register
            mov   [di]+2,cx    ; so transfer
            mov   [di],ax      ; to destination
            ret
;
cont:       sub   cl,dl        ; take Accumulator exponent
                               ; from Register exponent
```

```
            jz      sign          ; if they are equal deal with signs
            ja      do_acc        ; if Register has larger exponent
                                  ; shift the Accumulator
            neg     cl            ; otherwise make count positive
            cmp     cl,0eh        ; see if binary point to move
                                  ; more than 14 places
            jbe     shift_reg     ; if so, Register number too
            ret                   ; small and result already
                                  ; in Accumulator
shift_reg:  shr     ax,cl         ; adjust point in Register
            jmp     sign          ; and keep on
;
do_acc:     cmp     cl,0eh        ; check if Accumulator
            jbe     shift_acc     ; negligible
            mov     [di],ax       ; if so, answer is Register
            mov     cx,[si]+2     ; transfer it
            mov     [di]+2,cx     ; to Accumulator
            ret                   ; and leave
;
shift_acc:  shr     bx,cl         ; adjust point in Accumulator
            mov     cx,[si]+2     ; get exponent of answer
            mov     dl,cl         ; (same as for Register) in position
;
sign:       cmp     ch,dh         ; see if numbers have same sign
            jz      ok_add        ; if so, go ahead
            neg     ax            ; otherwise change sign
                                  ; of register
ok_add:     add     bx,ax         ; do addition
            jc      complete      ; if CF=1, finished
            cmp     ch,dh         ; otherwise, see if numbers of
                                  ; same or different sign added
            jz      complete      ; if same, done
            neg     bx            ; otherwise, make number positive
            xor     dh,1          ; and reverse sign for Accumulator
;
complete:   call    adjust        ; put number in standard format
            mov     [di],bx       ; and store
            mov     [di]+2,dx     ; in Accumulator
            ret                   ; and finish
;
;
adjust:     test    bx,8000h      ; see if 1 in bit 15
            jnz     large         ; if so, large number
;
```

```
pointl:     test   bx,4000h    ; see if 1 in bit 14
            jnz    exit        ; if so, format OK and leave
            shl    bx,1        ; otherwise, shift number left
            dec    dl          ; and reduce exponent by 1
            jnz    pointl      ; if no underflow, round again
            mov    bx, 0       ; with underflow put
            mov    dx, 0       ; everything to zero
            ret                ; and leave
;
large:      shr    bx,1        ; move number to right
            inc    dl          ; and increase exponent by 1
            cmp    dl,0ffh     ; check for exponent overflow
            jnz    exit        ; if none, finished
            mov    bx,0        ; otherwise, expunge BX
exit:       ret                ; return to caller
```

The program commences, as does the one for multiplication, by considering what happens when one of the exponents has the special value 0. It is always sensible to begin with arrangements for the non-normal or pathological cases that may arise. So the opening instructions should separate off the pathological from the normal. Whether you deal with the unusual immediately or postpone action till later is then a matter of taste and context—there is no universal rule to cover all circumstances. Here they have been tackled immediately to prevent the machine going through a long rigmarole before undertaking a simple operation.

You will notice that the program assumes that neither exponent originally in the Floating Point Accumulator and Register is 255 i.e. that the programmer has ensured that this pathological case is eliminated. This attitude is probably too trusting. Even in the absence of human error, previous operations in arriving at the numbers to be placed in the Register and Accumulator may have produced exponent overflow in one or other.

Generally speaking, in interfacing programs and modules one should fear the worst and expect that some mischance will feed input not in the specified range. So, instructions should be included to cope with such a possibility though they may be nothing more than a signal rejecting the input.

After the exponents of the numbers have been brought into conformity the question of sign has to be handled. The reason is that the sign is detached and kept in a separate location from the bodies of the numbers to be added. If the two signs are the same the addition is straightforward and, since the larger number must have a 0 in bit 15, there is no overflow or carry attached to the sum. When the signs are opposite we must make one number negative before the addition. We have settled on reversing AX, i.e. the Register, so that the Accumulator is always regarded as positive. So, if the Register and Accumulator were initially m_1 and m_2 we actually add $2^{16}-m_1$ and m_2 because changing sign is done by twos-complement. Thus, if $m_2 > m_1$, the result obtained is $m_2 - m_1$ with a carry and the appropriate sign is that of the Accumulator. On the other hand, if $m_2 < m_1$, the addition supplies $2^{16} - m_1 + m_2$ without a carry; here, to recover the correct format, we take a twos-complement, getting $m_1 - m_2$, and then the 1 in DH must be altered to 0 or vice versa to keep the record straight.

As a consequence OK_ADD first checks the Carry Flag CF. If CF = 1, this corresponds to $m_2 > m_1$ and nothing further need be done. If CF = 0 we first see if the original signs are the same and, if they are, the end has been reached again. In the contrary case, we have $m_2 < m_1$ and the manoeuvres described at the end of the preceding paragraph are gone through.

The sum achieved in this way may not have its leading 1 in bit 14. The final step, therefore, moves the 1 to bit 14 while making appropriate alterations to the exponent. This is the purpose of the module ADJUST which draws its input from BX,DX and returns its output there. It contains safeguards against underflow or overflow of the exponent. The module could, naturally, be called by other programs wanting numbers in the correct format provided that the proper information is furnished in BX and DX as input.

Exercises

For exercises 2–5, 9, 10, and 16 construct assembly programs for the processes described

1. Write a program to achieve SWITCH using XOR only without NOT.
2. Set the Carry Flag. Make a copy of bits 0–7 of the flag register in the byte at FLAG and then change ZF to its ones-complement.
3. Count the number of bits which are 0 in AX, placing the result in CX.
4. The number in BX is to be rounded to a byte in BH. Write a suitable module if BX is (i) an unsigned magnitude (ii) signed number.
5. Count the number of 1s in BX by employing SHR and observing CF after each shift.
6. Could ASC_BCD be rewritten so as to use LOOP for its cycling?
7. Rewrite ASC_BCD so that it complies with the Intel convention for the storage of packed BCD.
8. Eight ASCII bytes are stored in four bytes of packed BCD at HOME. Write a program BCD_ASC to recover the eight ASCII bytes, placing them at BASE.
9. Eight 4-digit numbers are stored in packed BCD at ARRAY. Add them and place the sum in packed BCD at ANSWER, allowing for the result occupying more than 4 digits.
10. Two 4-digit numbers are stored in unpacked BCD at ALPHA and BETA. Multiply them and put the result in unpacked BCD at GAMMA.
11. Convert the following numbers to binary floating point and then display

them as they would appear in the Floating Point Register (use binary or hexadecimal notation according to your preference).
(a) 2.0, (b) 32.375, (c) -32.75, (d) 0.027×10^3, (e) 625×10^{-3}, (f) -12.8×10^2.
12. The bytes in the Floating Point Accumulator are shown (in hexadecimal). What decimal numbers do they represent?
(a) 48 00 00 80, (b) 70 00 00 68, (c) 78 00 01 8A.
13. Insert in FLPT_MUL instructions to (a) preserve registers used, (b) to cope with rounding.
14. Insert in FLPT_ADD instructions to check that initially the Accumulator and Register are not indicating overflow.
15. An unsigned magnitude occupies the word at INTEGER. Transfer it to the Floating Point Accumulator in correct format.
16. (continuation). A second unsigned magnitude is in the word at ANOTHER. Transfer it to the Floating Point Register and then add the two numbers. Would you obtain the same result if you added INTEGER and ANOTHER directly, then converted the result to the Floating Point Accumulator?
17. Consider Exercises 15 and 16 for signed numbers.
18. The coefficients A,B,C, in Ax^2+Bx+C are integers in consecutive words at COEFF. It is known that the quadratic has opposite signs at $x=1$ and $x=8$. Find, in floating point arithmetic, the value of x where the quadratic vanishes by the *bisection method*. In the bisection method, if the quadratic has opposite signs at $x=a$ and $x=b$ $(b>a)$, the value at $x=c=\frac{1}{2}(a+b)$ is calculated; if the result is zero then $x=c$ is the desired answer; if the value at c has the opposite sign to that at a repeat the procedure for the interval (a,c) otherwise repeat for (c,b). (This is an example of a *binary search*.)
19. Consider generalizing Exercise 18 to finding a real x for which $A_nx^n + A_{n-1}x^{n-1} + \ldots + A_0$ is zero, assuming that $n, A_0, \ldots, A_n$ are available in memory in floating point format.
20. Draw up a program for floating point division.

7
INTERRUPTS

7.1 Introduction

From time to time a microcomputer has to accept signals from the outside world and despatch messages to others if it is to serve a useful function. Such activity will normally interfere with its usual flow and, during it, the processor is said to be handling an *interrupt*. It is desirable that the program shall be able to stop at an appropriate instruction, cope with the interrupt and then return to its original duty without losing any information. In general, interrupts arrive at unpredictable and often inconvenient times so that the program must be able to maintain its integrity while dealing with them. The timing of the pressing of a key on the keyboard will rarely coincide with the clock controlling the processor. Thus, when a key is pressed, there have to be arrangements whereby either the character is accepted at once or it is held in a *buffer* to await later treatment. The two options correspond to answering a telephone as soon as it rings or storing a message in an answering machine until a convenient moment. In brief, then, an interrupt is an *asynchronous event* which has to be handled by the processor.

The interrupts so far described are essentially external in nature. There is also the possibility that we may wish to incorporate them deliberately in our program, perhaps to print a file; such *internal interrupts* will be considered more fully in Chapter 9.

Basically, there are two methods of dealing with interrupts—via hardware and via software. For instance, the input might be connected directly to the memory without passing through the processor. Data can be transferred very rapidly, limited only by the speed of the input or memory, whichever is slower. So *Direct Memory Access* (or DMA for short) is valuable for high-speed communication over permanent lines. The price that is paid is in the provision of a special bit of hardware, the *DMA controller*, which implements the route between the input and memory. Nothing further will be said about implementation by means of hardware; we shall concentrate on techniques based on software.

7.2 Polling

When an input device has some information for the processor there are two ways in which this can be discovered. Either the input sends an alerting signal or the processor checks periodically whether the input has anything to deliver.

Analogously, these correspond to opening a door only when the bell rings and going to the door at regular intervals to see if there is a caller. The second of these methods, in which the program systematically *polls* the input, will be discussed in this section. In the next section we shall study the other method.

The advantage of polling is immediately evident. Control lies entirely with the program, which does not have to pay any attention to the input until it wishes to do so. On the other hand, polling is also pretty inefficient because, if an input is not to be missed, the input must be checked frequently and, on most occasions, there will be nothing for the program to receive. Polling is therefore a device for when program control is essential or the program does not have many other duties to occupy its time. Generally speaking, it is for the low-speed transfer of data.

A simple example of a polling program will now be given. In this example, the input changes bit 7, the sign bit, of the byte declared at ALERT from time to time. We wish to know on how many occasions the input set the marker to 1. Clearly, our program will have to poll the bit 7, do nothing if it finds 0 and increase the counter if it detects 1. But that raises a question which often has to be faced in polling. If a 1 has been detected and, on the next interrogation, a 1 is discovered is it a new 1 or the old 1 which has not yet disappeared? One way round this difficulty is for the program to tell the input that the 1 has been recorded and the marker can be switched off. When that avenue is not open, as will be assumed here, the safe, if tedious, action is to keep on polling (without counting) until 0 appears.

In the following fragment the counting is done in CX. It is assumed that this has been started at 0 at a suitable stage in the main program.

```
poll:   test    alert,80h       ; test the byte with a
                                ; logical AND
        jns     poll            ; if sign bit 0, poll again
        inc     cx              ; otherwise, add 1 to the
                                ; counter
stay:   test    alert,80h       ; test the byte again
        js      stay            ; if SF=1 don't leave
        jmp     poll            ; ready for the next 1
```

There are two points to observe. Firstly, we have not put BYTE PTR ALERT because of the earlier assertion that ALERT has already been declared to be a byte so the assembler does not require repetition of this information. Secondly, the two jumps to POLL are made for simplicity. In practice, you would want to carry out other instructions between interrogations of the bit 7 and so you would arrange to jump back to some convenient label in the main program, which would have the responsibility of returning to the polling instructions sufficiently frequently that you did not fail to catch a 1. In contrast, the STAY after JS is firm so as to prevent any further counting until certain that the 1 has been terminated.

7.3 Interrupt Service Routines

We now turn to the method in which the program does not poll but follows its own instructions until told that its service is requested by an interrupt. A special interrupt path or line into the processor has to be provided for this purpose. When the processor is alerted to an interrupt demanding attention it first completes its current instruction.

Its next step is to examine the *priority* of the interrupt. If the priority is no higher than that of what is already being processed, the processor ignores the interrupt and completes the task in hand. Only then will it consider accepting the interrupt. Similarly, the processor refuses to begin at once on an interrupt if its priority (and lower levels) have been *masked* to prevent processing. In other words, an interrupt has to have higher priority than what is being handled already before the processor attends to it.

The 8086 has a special line (NMI) for a top priority interrupt. An interrupt on this line cannot be masked and is always accepted by the processor after the current instruction is finished. It is therefore a *nonmaskable interrupt*—which explains the name of the line. Such an interrupt could serve as a warning of an impending power failure so that the processor has an opportunity to take protective measures.

Once the processor accepts the interrupt, control goes to a special program, the *Interrupt Service Routine*, which deals with the interrupt. Its job, apart from its functions for the data from interrupt, is to see that control goes back to the program that was running when the interrupt occurred and that the program picks up from where it left.

At some stage in the proceedings the processor should inform the sender of the interrupt that the interrupt has been accepted. This *acknowledgement* that communication between two devices is going forward is sometimes referred to as *handshaking*.

Most of the above actions are undertaken automatically by the processor. What has to be supplied is the interrupt service routine and its address, so that the processor knows where to transfer control to. The address of the first instruction of the interrupt service routine is known as the *interrupt vector*. It is usually in a special table, called the *vector table*, since several different kinds of interrupts will have to be managed normally. The interrupt vector fills a doubleword of memory since both segment and offset have to be stored, the offset being in the first word and the segment in the second. For this reason, a special instruction LDS is available for retrieving vectors from memory. Thus

```
lds     dx, intvector
```

transfers the first word of INTVECTOR to DX and the second word to DS. (A similar instruction LES uses ES instead of DS). To avoid undue manipulation of double words the entries in the vector table are identified by number. Restricting the number to a byte permits 256 entries which should be adequate in many circumstances and has the benefit that only a byte has to be manoeuvred.

The vector number may be determined in two ways. Either the vector number is despatched by the interrupter shortly after it indicates an interrupt or the priority level signifies a particular number. In the second case an *autovector* is said to be selected. For example, an interrupt on the NMI informs the processor that it should go to vector number 2 without any more ado i.e. an autovector has been employed. The NMI interrupt does not itself have to transmit the number 2 along the line.

7.4 Further details for the 8086

While much of what has been said in the previous section is applicable to any 16-bit processor, some further detail is necessary to an understanding of how the 8086 performs its functions in order to implement interrupt service routines.

The 8086 has three lines, restricted to input only, for interrupts. One of these, the NMI, has already been described. Another, entitled RESET, has absolute right of way because it stops the processor in its tracks and restarts execution in a pre-determined state (via read-only memory) without any reference to its earlier operations. Principally, it is for clearing hang-ups and starting the computer in a prescribed manner.

The third line, labelled INTR (short for *Interrupt Request*), is for the more conventional interrupts. It interrupts the processor when it is at a high level but not at a low level. Masking can be accomplished by the Interrupt Enable Flag IF. If IF=0 no notice is taken of what happens on INTR. If IF=1 an interrupt occurs when INTR is raised; likewise, if INTR is high and then IF is set to 1 the processor is interrupted. The instructions CLI and STI clear and raise the IF flag respectively.

In order to be aware of the state of INTR the processor checks it after the execution of each instruction (except those setting segment registers). You might regard this as concealed polling but it is inevitable if an interrupt is not to be missed. The exception mentioned at the end of the first sentence of this paragraph discloses a noteworthy feature of the 8086. After a MOV or POP alters a segment register interrupts are masked or disabled for one instruction. The reason for this arrangement is to permit a sequence such as

```
mov     ss,ax
mov     sp,bx
```

to proceed without fear of interruption. If INTR was interrogated after the execution of the first of these two instructions then an interrupt could be accepted immediately. It would then arrive with the stack undefined because the offset of the stack pointer, carried by SP, would not have been properly adjusted. The consequence would be extremely peculiar. Since an instruction modifying a segment register may be followed by one adjusting the relevant offset the 8086 affords protection by not looking at INTR at the end of the execution of an instruction changing the content of a segment register or, rather, by actually disabling interrupts. Instructions concerning the offset, e.g. the second one above, are excluded from this rule, which refers solely to segment registers, so that after SP had been changed above INTR would be sampled.

The processor acknowledges receipt of a request on a path known as INTA (*Interrupt Acknowledge*). After satisfactory acknowledgement the interrupter places the vector number on INTA (there are no autovectors with INTR so an interrupt has to furnish a vector number) so that the processor can find the interrupt vector. The processing also clears the Interrupt Enable Flag to mask any future interrupts until it is reset.

All that remains for the processor to do before invoking the Interrupt Service Routine is to preserve enough information to be able to go back to its original program eventually. This is does by placing certain registers on the stack (the saving of other registers is the responsibility of the Interrupt Service Routine). These records have to be recovered on return and so a special instruction IRET is used as the exit from an Interrupt Service Routine.

A summary of the steps involved is:

1. Request for interrupt.
2. Processor completes current instruction.
3. If NMI, put vector number to 2 and go to step 10.
4. If INTR and IF=0, pay no attention.
5. Otherwise acknowledge on INTA.
6. Acknowledge again on INTA.
7. Requester provides vector number on INTA.
8. Processor reads vector number.
9. Requester releases data lines.
10. Push flags register on stack.
11. Push CS on stack.
12. Push instruction pointer IP on stack.
13. Load CS and IP indicated by vector number.
14. Transfer control to Interrupt Service Routine.

Notice that all these steps are carried out without any intervention by the programmer whose sole concern is the Interrupt Service Routine. The procedure uses three levels of stack so that IRET must, on its own accord, pop these three in the correct order on the termination of the interrupt before the interrupted program can continue.

Note that when a request for an interrupt is issued the processor first leaves it alone until it has dealt with the current instruction. After that its first duty is to confirm if it is on NMI; if so, it accepts the interrupt and moves immediately to step 10 to preserve some registers followed by entering the Interrupt Service

Routine. If the interrupt is not NMI but on INTR the processor checks the Interrupt Enable Flag. If it is cleared the interrupt is passed over and the processor carries on with its previous activities.

If, however, IF = 1 the processor prepares to accept the interrupt and sends an acknowledgement (twice). The requester then provides the byte for the vector number and, once the processor has read this, then releases the data bus. Thereafter, the processor proceeds to the stacking of registers and invoking the Interrupt Service Routine.

As one of the simplest illustrations consider counting the number of interrupts in CX. The matter of setting the interrupt vector will be omitted since it is more conveniently discussed in Chapter 9.

```
begin: hlt          ; wait until interrupt occurs
       inc   cx     ; count it
       iret         ; back to interrupted program
```

This primitive program has the processor sitting around waiting for an interrupt. When the interrupt occurs it adds 1 to the count and then returns to its sedentary occupation.

If the instruction HLT were not available a similar result could be achieved by writing

```
begin: jmp   begin
```

When the program brings JMP from memory the Instruction Pointer is pointing to the next instruction. The effect of the JMP is to alter the IP to point again to the JMP instruction so that the program is to bring JMP from memory. Thus, the processor is essentially doing nothing.

At the intervention of an interrupt the current instruction is completed. When it is over the IP will be pointing to JMP. This is saved on the stack before entering the Interrupt Service Routine and then recovered (via IRET) on return so that the processor continues in its waiting state.

The Interrupt Enable Flag referred to above is a control flag. When it is raised external interrupts are enabled i.e. the processor recognises signals coming into the computer from outside sources such as the keyboard, printer, etc. and handles their requests for service. When IF is cleared, no attention is paid to these requests. There are instructions CLI, to clear or disable interrupts, and STI, to set or enable interrupts, available for you in your programs. However, the effect of the operating system cannot be ignored. Many input/output Interrupt Service Routines themselves disable external interrupts and then enable them before returning control. If you are having difficulty with keyboard or printer transfers it may be helpful to enable interrupts at the beginning of your program, though this should not be necessary with the system calls of Chapter 9.

7.5 Input and output

Getting a character from the keyboard into the processor entails an interrupt. Sending a character from the processor to the screen or to a disk file likewise goes through an interrupt. A similar situation arises in sending data to or receiving information from another computer. In fact, many Interrupt Service Routines are concerned with accepting characters from one place, doing something with them and transferring the results to another location (see also Section 9.1). In other words, the standard properties which the microcomputer user expects of input and output are making constant use of interrupts. A further facility the user expects is a device for the updating of the time and calendar (often attached to files) and here again interrupts are involved. Our examples will therefore deal with these common interrupts without implying that there are no other important interrupts.

Implementation via DMA might be pertinent in some cases but that possibility will be omitted here. External connection is via *ports* which are numbered in the range 0–255. The number associated with a port for input or output is dictated by hardware considerations. Without precise knowledge of the hardware it is impossible to specify exactly what those numbers will be, so we shall assign them arbitrarily.

The first example comes from a statistical calculation during which some scores in the range 0–9 are to be stored, as well as their sum and the number of times that 0 was awarded. It is assumed that there are not more than 255 scores. The digits are in ASCII bytes and enter via port 10. The main program may well be referring to the input from time to time so it is supposed that it has a declaration such as

```
sum          rw   1
count        rb   1
zero_count   rb   1
number       rb   100h
```

somewhere and also that it has initially set the sum and counts to zero. Then the Interrupt Service Routine could be

```
stat_int  cseg                ; start of code segment
          org    1000h        ; leave plenty of room for
                              ; main program
open:     push   ax           ; save registers
          push   bx           ; used by Routine
;
          xor    ax,ax        ; clear AX
          xor    bh,bh        ; clear BH
          mov    bl,count     ; and put count in BX
```

```
;
            in      al,port              ; collect byte from input
            sub     al,30h               ; remove ASCII offset
            js      over                 ; if result negative, not a digit
                                         ; so ignore
            jnz     illegal              ; if not zero, see if illegal
            inc     zero_count           ; otherwise add 1 to count
                                         ; of zeros
  illegal:  cmp     al,9                 ; see if greater than 9
            ja      over                 ; if so, illegal and ignore
            add     sum,ax               ; otherwise adjust sum
            mov     number[bx],al        ; store digit
            inc     count                ; alter the count
;
  over:     pop     bx                   ; restore registers
            pop     ax
            iret                         ; return to main program
;
            dseg                         ; start of data segment
  extrn     sum: word,number:byte        ; declared
  extrn     count:byte,zero_count:byte   ; elsewhere
  port      equ     10                   ; number of port
```

Three points are worthy of remark. The first is that we have tried to make the routine a little less like a fragment by including some of the assembler directives that have been mentioned earlier. Second, the two registers AX and BX which are used by the routine are saved at the beginning and restored at the end. Third, an error checking mechanism has been introduced to ensure that only genuine digits are processed. The ASCII representation for digits is 30H–39H so 30H is subtracted and then it is confirmed that the result lies in 0–9 before the store is affected.

When you are *absolutely* confident that comments can be dispensed with, multiple instructions can be written on a single physical line by separating them with exclamation marks e.g.

```
over:     pop bx! pop ax! iret
```

because the assembler treats the ! as an end-of-line marker. Remember, however, that only the last statement on a line can contain a comment; since the assembler regards a comment as running to the physical end of a line any instructions (with or without an exclamation mark) after the start of a comment will be deemed to be part of the comment and ignored as instructions.

7.6 *ASCII letters*

The next example takes ASCII letters from input port 10, converts them to upper case if necessary and then sends them to port 20. The output expects to receive its characters with odd parity (see Section 8.3 for a possible reason why) and so this has to be arranged before despatch. This time a check that a genuine letter has come from the input will be excluded. Also the data segment is placed first so that you can form an opinion as to which structure you find preferable.

```
            dseg                 ; start of data segment
            org     1000h        ; leave room for main program
port        equ     10           ; specify input
output      equ     20           ; and output
;
            cseg                 ; start of code segment
letters:    push    ax           ; preserve register
            in      al,port      ; collect byte from input
            cmp     al,'a'       ; if ASCII less than 'a'
                                 ; must be upper case
            jl      uppercase
            sub     al,20h       ; otherwise convert to upper case
uppercase:  jpo     ready        ; if parity odd, OK to send
            or      al,80h       ; otherwise, put a 1 in bit 7
ready:      out     output,al    ; transmit byte to output
            pop     ax           ; restore register
            iret                 ; back to main program
```

Here the relevant information is that in ASCII each capital letter is 20H below its lower case counterpart. Also the representation for 'a' is lower than that for 'z'. Hence AL is compared with 'a', employing the notation for a string as in Chapter 4, and if found to be lower it must already be upper case. If it is lower case the subtraction of 20H transforms it to upper case. An alternative instruction that achieves the same object is

```
and     al,0dfh
```

since that guarantees a 0 in bit 5 which is what the subtraction of 20H also does.

The conditional jump JPO leaps over the next instruction if the parity is already odd. If the parity is even the OR inserts a 1 in the left-most bit of AL and makes the parity odd.

7.7 *Clock updating*

The problem of updating the clock and calendar is rather more recondite. For simplicity only the clock will be discussed. It will be assumed that the main program has already declared bytes at HOUR, MINUTE, SECOND and given them the correct starting values. Also it has declared a byte at CNT with the value 100 for counting purposes. It then issues an interrupt every 1/100 second for the Interrupt Service Routine to keep the clock on time. The hour is supposed to be recorded in the standard 24 hour style.

The program contains three portions which could be similar in structure—one for seconds, one for minutes, and one for hours. There could therefore be certain repetition of instructions. In order to introduce variety and bring out a point two different methods of formulating the instructions have been deployed. However, further explanation will be deferred until you have had a chance to look at the program.

```
            dseg                        ; start of data segment
   extrn    second:byte,minute:byte
   extrn    hour:byte,cnt:byte
   max_cnt  equ      100                ; constant for resetting
                                        ; counter
            cseg                        ; start of code segment
   clock:   dec      cnt                ; reduce counter by 1 at
                                        ; each interrupt
            jz       upsec              ; if counter down to 0, time
                                        ; to change seconds
            iret                        ; otherwise, back to main program
;
   upsec:   push     ax                 ; preserve registers
            push     si                 ; on stack
;
            mov      al,max_cnt         ; reset the counter
            mov      cnt,al
;
            lea      si,second          ; address of SECOND in SI
            inc      byte ptr[si]       ; add 1 to second count
            cmp      byte ptr[si],60    ; is it up to 60?
            jne      finish             ; if not, finished
;
            mov      byte ptr[si],0     ; if so, clear the seconds
                                        ; and deal with minutes
   upmin:   inc      minute             ; add 1 to minutes
            cmp      minute,60          ; is it up to 60?
            jne      finish             ; if not, done
;
            mov      minute,0           ; if so, reset the minutes
                                        ; and deal with hours
   uphour:  inc      hour               ; add 1 to hours
            cmp      hour,24            ; is it up to 24?
```

```
            jne     finish          ; if not, finished
            mov     hour,0          ; if so, reset
;
finish:     pop     si              ; restore
            pop     ax              ; registers
            iret                    ; return to main program
```

You may be surprised, in the opening instructions under CLOCK, at the absence of any saving of registers on the stack. The reason is that the hundredths of a second are not being recorded, only their passing counted. Only when a full second has elapsed do we alter the record. Therefore for 99 of each 100 interrupts the Interrupt Service Routine can be very brief and will not involve any registers.

When we turn to the main part of the program relevant registers must be preserved initially and returned to their original state at the end.

Each of the sections for updating the second, minutes, and hours is based on the same principles. When appropriate the record is increased by 1. The result is then compared with the maximum permitted for that record. If that maximum has been reached we reset the record to zero and move to adjustment of the next in the sequence; otherwise all that is necessary is to transfer control to the closing instructions.

Although the principles are the same two different ways of implementing them are demonstrated in UPSEC and UPMIN. In UPSEC the address in SECOND is moved to SI and then SI acts as a pointer for subsequent instructions via indirect addressing. However, once the address is in SI the assembler does not know whether it is pointing to a byte or a word unless another operand in the instruction gives it the information. For instance, if it is told to copy AL to [SI], it is cognisant that a byte is the subject of the operation. Three instructions in UPSEC are not in this category. Either there is a single operand or the other operand is an immediate value offering no clue as to whether it is byte or a word. Therefore the assembler has to be instructed that a byte is pertinent by the insertion of BYTE PTR.

UPMIN addresses the memory directly. Since the assembler is already familiar with the fact that MINUTE has been declared as a byte it does not face the same ambiguity as in UPSEC and can dispense with a further reminder. Consequently, BYTE PTR is absent from UPMIN.

One instruction fewer is needed for UPMIN than UPSEC because an address does not have to be loaded into SI. Moreover, direct addressing generally occupies less time than indirect so that UPMIN will be processed faster than UPSEC. If UPSEC were brought into line with UPMIN the instructions saving and restoring SI could be omitted. While you might not think that saving a few microseconds is important you should realise that unnecessary delays in sets of instructions that are gone through frequently can soon accumulate into appreciable and undesirable amounts.

Exercises

8–10 are exercises for which you should provide Interrupt Service Routines

1. Modify POLL so that, after detection of a 1, it switches the marker off (assuming the input permits this).
2. How many handshakes occur in the steps in Section 7.4?
3. Modify LETTERS so that it ignores any characters other than letters in the input.
4. Modify LETTERS so that, in addition to transmitting letters, it passes on punctuation unchanged but ignores any other characters.
5. Modify LETTERS so that it despatches its characters to the output with even parity.
6. Port 35 provides a byte at each interrupt. Write an Interrupt Service Routine which copies it to the address in SI and increments SI.
7. Repeat Exercise 6 when the address is to be found in memory location 1234H.
8. When a byte arrives in port 36 send it to the next element in the array POS if it is positive, to the next element of the array NEG if it is negative and increment the word COUNT if it is zero.
9. In Exercise 8 accept no more input as soon as COUNT reaches 0FFFFH.
10. ASCII characters are in an array at HOME. At an interrupt send the next character to port 37 but omit any control character. Arrange to stop at the end-of-file character.
11. Add a section to CLOCK so that one-hundredths of a second are counted in HUNDRED.
12. Alter CLOCK to deal with interrupts which issue every 1/1000 second.
13. Append to CLOCK instructions for keeping the calendar up-to-date, assuming that each month has 30 days and every year 360 days.
14. Repeat Exercise 13 but with the proper calendar (do not forget leap years).

8
COMMUNICATION

8.1 Types of communication

Microprocessors are designed to work with bits but not all information comes in digitized form, e.g. books and speech. Accordingly, communication with a microprocessor may entail conversion of information of one type to digital format and vice versa. Books are relatively easy when in a standard alphabet since codes such as ASCII allow direct keying in. Speech is more difficult because it does not split up into discrete levels but varies continuously. Nevertheless, there are devices (analogue/digital converters) which transform speech and other continuous signals as well as those (digital/analogue converters) which carry out the reverse process, and some microcomputers are supplied with them. Books on information theory tell us that such conversions are bound to lose information even in the absence of noise and also indicate how to keep the loss of content within practical limits.

The theory of such devices is outside the scope of this book despite their significance for human enlightenment. Our survey has the less ambitious objective of enquiring what happens in the connection of the processor to a source or receiver of digital data. This sort of consideration arises when the processor sends output to a printer or is exchanging ideas with a remote terminal.

Such transfers may be classified into two categories—*parallel* and *serial.* In parallel transmission several parallel paths join source and destination so that bits can travel down them simultaneously. For example, to send a byte you could have eight parallel lines and despatch all the bits of the byte at the same moment, each bit assigned to its own line. So long as the lines are the same length the bits will arrive together at the receiver ready for recombination into a byte. Parallel communication is therefore fast and a common feature within processors. However, as the distance between source and destination grows the expense of catering for parallel transmission becomes daunting partly because of the cost of the provision of multiple lines and partly because of measures to ensure that the bits arrive in step without being unduly affected by noise.

One bit at a time is the rule for serial transmission. There is only one channel and the bits journey along it one after the other. This stream of bits resembles the flow of water from a tap whereas parallel transmission corresponds to the water descending Niagara Falls. Serial transmission is inherently slower than parallel but the cost of the connecting link is much cheaper. So parallel circuits tend to be employed for fast communication between devices not too far apart while serial signals are the custom where speed of transmission does not weigh against the costs of transmission.

8.2 Serial connections

Serial transmission is simple in principle—just send one bit after another—but it is quite easy for devices to be unable to communicate with each other unless certain standards or *protocols* are observed.

Before we discuss the details of the signals themselves let us look at the physical connections. A connection might be a cable joining two computers directly or it might be a rather more complicated link, as in Fig. 8.1, involving acoustic couplers or modems and telephone lines. The modem or acoustic coupler has the duty of making a suitable transition between the processor and telephone line. The protocols for the cables are pretty well accepted; at least there are standard plugs such as the RS232C. The pin attributes are also tolerably standardized though it is always wise to check in any particular case. Whatever the convention adopted the cable should have at least eight lines for the following functions—the *frame ground* will also be connected. The abbreviations in parentheses are sometimes met instead of the fuller titles.

Request to send (RTS). Tells the other device that I am ready to send data.
Clear to send (CTS). Indicates that other device is ready to accept data and enables transmission. Often the RTS and CTS pins are attached to a common wire.
Transmit data (TXD). For sending the data as soon as CTS indicates that the other device is ready.
Data carrier detect (DCD). For receiving the signal from the other device that it wishes to send data.
Data terminal ready (DTR). Signal to the other device that I am ready to accept data.
Receive data (RXD). For the reception of data from the other device.
Data set ready (DSR). Indicates other device ready for operation and prepared to accept control signals.
Ground (GND). A return line for signals.

Codes to assist with the performance of these tasks are among the ASCII characters 00H–1FH.

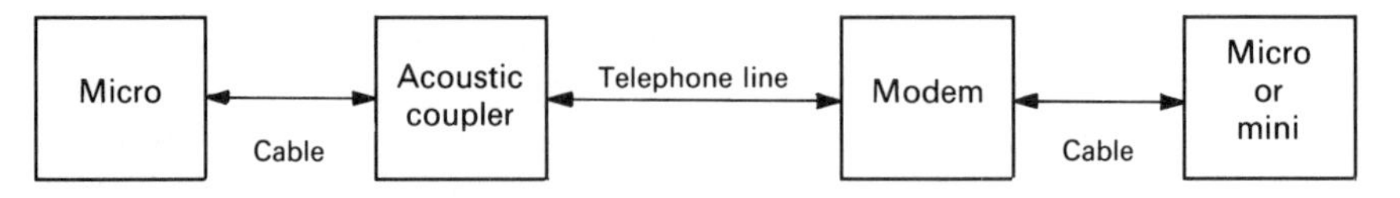

Fig. 8.1 A possible transmission circuit

It should be emphasized that the label at one end of the line will not be the same as that at the other. For instance, transmission at one end is reception at the other so that TXD goes to RXD. Similarly RTS at one end goes to DCD at the other and DTR to DSR.

The essence of these control signals is to ensure that the transmitter does not send any information to the recipient until it is acceptable but transmits as soon as allowed to do so. Often the tempo is improved by having two buffers for reception (and two for transmission). The first buffer takes the stream of bits. As soon as it can assemble a character it transfers the character (by parallel transmission) to the second buffer where it is held until the processor reads it. This mechanism prevents interruption of the stream of bits while awaiting the processor's reading of the whole character. The transmission buffers undertake the opposite procedure.

8.3 Conventions

Characters are transmitted a bit at a time in serial communication. The speed of transmission is measured in *bauds*, 1 baud corresponding to 1 bit per second. Standard rates that have been settled on are 75, 110, 134.5, 150, 300, 800, 1200, 2400, 4800, 9600, 19200 baud (disk transfers often take place at much higher rates like 38400 baud). Normally, the rates of transmission and reception are the same but some video text databases use 75 baud for transmission and 1200 baud for reception; commonly, this option on a microcomputer is accompanied by the reverse 1200 baud for transmission and 75 baud for reception.

Some concept of these speeds is obtained by observing that at 110 baud a bit is arriving about every hundredth of a second (every 9 milliseconds is closer) and, at 19200 baud, every 52 microseconds. To put it another way, if an average English word needs 60 bits for reliable transmission, 110 words of English text will be sent per minute at 110 baud and 19200 words (say 60 pages of text) in a minute at 19200 baud.

Clearly, in order for one device to 'talk' to another both must be able to operate at an agreed speed of transmission and the intervening link must be capable of sustaining this speed with accuracy. Verification that both machines are adopting the same parameters is sometimes termed *matching the handshaking.*

Some control protocols have to be agreed also. For instance, there may be the facility for choosing to keep or dispense with DTR, DSR, RTS, and DCD; without an agreement on their presence or absence control signals and text may become confused. A related communication control is the protocol for XON/XOFF, an abbreviation for Send ON/Send OFF. The purpose of this protocol is to prevent overflow. It has already been mentioned that data will be temporarily stored in a buffer. If the speed of transmission is greater than the speed with which data can be unloaded from the buffer incoming data will be lost on account of buffer overflow. Therefore, when XON is selected, the receiver sends out a signal such as 13H when its buffer is, say, three-quarters full. This warns the transmitter to stop sending data until the receiver is ready to handle further information. When the content of the receiver has dropped to, say, one quarter it sends out another control signal such as 11H telling the transmitter to begin sending again. In this way buffer overflow is avoided. Care must be taken in opting for XON if machine code, as opposed to ASCII, is being transmitted. For, if the machine code contains the XON or XOFF characters, they will be acted upon and lost from the machine code file. Not only that, but the effect may be that the transmission hangs indefinitely. For this reason the usual default setting is XOFF in which no action is taken.

Other features have to be considered in matching the handshaking. Obviously the code to be adopted for a character is one. The normal one is the ASCII code, though any other which does not require more than a byte is all right so long as it is acceptable to all concerned. In fact, the restriction to a byte is not really necessary as far as principles go but it is commonly adopted. It will be assumed in the following that it has been agreed to send ASCII characters.

The next decision is wheter to use 7 or 8 bits for transmitting an ASCII character. Of course, if you want to send 256 different characters there is no alternative to deploying the full 8 bits. But you may desire only those characters in the ASCII list up to 127. Then 7 bits conveys enough information; all that has to be done by the recipient is to add a 0 as the most significant bit to recover the complete byte (bit 7 of a byte is often designated the *most significant bit* or MSB, for short). Thus the rate of transmission of characters can be speeded up by a worthwhile factor.

Notwithstanding this, you may still prefer to work with 8 bits with the MSB serving for error checking. The simplest and weakest form of this is to confirm that the MSB is not a 1 otherwise an error is certain to have occurred. A better check, though not all that strong, is to employ the MSB as a parity bit. Then both transmitter and receiver must be fixed at the same parity either odd or even. Suppose it is even. Then the 8 bits must always contain an even number of 1s, arranged by adjusting the MSB so that A is sent as 0100 0001 and C as 1100 0011. If the receiver finds the parity is even it accepts the character. If the parity is odd it still accepts the character but raises an error flag to warn that a parity error has beeen detected. Naturally, a double error can get past this guard. For example, if C were transmitted with parity and received as 1010 0011, no error flag would be raised but the character would be interpreted as #. In reasonable lines, two errors in 8 bits are unlikely so that parity checking can be quite effective. What is more probable is that outside interference—power surges or other electrical disturbances—will be on a longer time scale than that of data transmission so that groups of characters tend to be spoiled rather than single bits. On these grounds some argue that parity checking scarcely merits the bother.

Although it was stated earlier that 7-bit transmission limited one to half the ASCII list this is not true if a Shift-In/Shift-Out (or SI/SO) facility is available.

This facility, which applies only to 7 bit transmission, may be either on or off. When off no action is taken i.e. the MSB of the byte is understood to be 0. In the on position, the reception of an SI character (0FH) will entail all subsequent characters having a 0 added as MSB to complete the byte. If the SO character (0EH) is received, subsequent characters in the range 20H to 7EH have 1 added as MSB to finish the byte; characters 00H to 1FH and 7FH are given 0 as MSB. By this means the range of characters which can be sent is considerably extended. Again, there may be danger as with XON/XOFF when machine code is being received and SI/SO is on. If the SI or SO characters are part of the machine code they will be implemented and lost to the machine code file. Moreover, if SO is received, the data following will be altered by having the high bit set causing additional corruption to the file. Consequently the default setting of SI/SO is off.

8.4 Synchronization

There remains the question of how the receiver knows where a character begins and ends. Basically, there are two methods of resolving this question.

The first involves *synchronous* signals which, in essence, means that transmitter and receiver are run by the same clock. Timing rules for the transmission of each bit and character are laid down. By referring to the common clock the receiver can determine starts and finishes thereby being enabled to decipher the signals from the transmitter. Keeping the timing right for both transmitter and receiver is non-trivial and synchronous systems therefore tend to be expensive. The ordinary microcomputer user is therefore less likely to encounter synchronous communication than the next method.

The second method, called *asynchronous* to distinguish it from the first, demands only that the transmitter and receiver clocks march together sufficiently closely for the period that a character lasts. Then all that the transmitter has to do is to send an opening signal telling the receiver to start its clock, transmit the character and then despatch a closing signal. Thus, each character carries a synchronizing marker and the two clocks need only be kept together for the duration of the bits of the character.

Let us see how this is achieved. The channel connecting the transmitter and receiver can support two levels of signal, designated 1 and 0 respectively. In the idle state the channel is maintained at level 1 (sometimes called the *mark level* in serial communication). At the beginning of a character the transmitter changes the level to 0 (sometimes known as the *space level*) to indicate to the receiver that it should start its clock running. This *Start Bit* is the synchronizing marker that enables the receiver to keep in step with the transmitter character by character.

It is customary to send the bits beginning with the least significant and proceeding to the most significant. Hence to send F (0100 0110) the signal displayed in Fig. 8.2 will appear on the line. All of the bits are of the same length

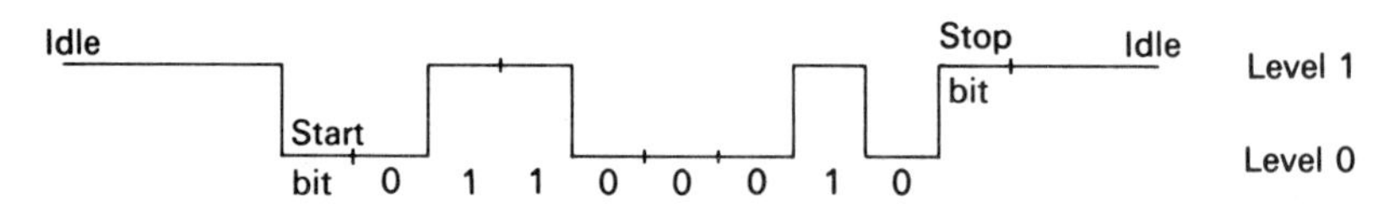

FIG. 8.2 Transmission of F

so that the receiver is aware of the position of each successive bit. At the end of the character the transmitter returns the line to level 1 thereby signalling a *Stop Bit*. Thereafter the line stays at level 1 until the initiation of another Start Bit.

Fig. 8.2 shows a single Stop Bit but often an option is provided to prolong the stop to 2 Stop Bits or even 1.5. (According to convention 2 stop bits are reserved for 110 baud lines and 1.5 to transmission of 5 bit Baudot code). It is transparent that, before transmission commences, transmitter and receiver must be in harmony as to the number of Stop Bits to be adopted in order to match handshaking. To fix ideas we shall assume that the option of 1 Stop Bit has been selected.

Until a Start Bit is received the receiver takes no action. If no Stop Bit is read in the correct position the receiver knows something has gone wrong but cannot reject the bits already accepted. Consequently, it raises an error flag reporting that a *framing error* has been met i.e. the character was not properly between the frames of the Start Bit and Stop Bit.

The task of the receiver in identifying the signal on the line is principally that of determining at suitable times whether the level is 1 or 0. The times should not be too close to the ends of a bit otherwise jitter might result in incorrect reading at changes between 0 and 1. Furthermore, brief transitions due to external sources might lead to spurious readings. Therefore, the sampling is usually done at a time when the signal should be firm and steady like the centre of the bit. The additional advantage stemming from such a choice is that the clock can get a little out of kilter without generating an error.

The clock controlling the timing, which will be called the *baud clock*, can be run at the same rate as the line. More flexibility is achieved, however, by running it faster, typical factors being 16 and 64. If the factor is 64 and the speed of transmission is 110 baud the frequency of the baud clock would be 7040 Hz; at 19200 baud the frequency would be about 1.2 MHz.

The sharp drop from 1 to 0 in the Start Bit causes the baud clock to begin running. After 32 cycles (assuming the clock is 64 times faster than the line rate) it checks the level of the line i.e. it tests at the middle of the Start Bit. If the level is 1 it concludes that it has made a false start and been triggered by a spurious signal; it reverts to its waiting state. On the other hand, if the level is discovered to be 0 the conclusion is that a genuine Start Bit has arrived and reception should commence. Naturally, neither conclusion can be guaranteed to be true but, given proper design, both can be asserted with high probability.

Confident that there is a signal to be accepted, the receiver tests the line level after a further 64 cycles (i.e. 96 cycles after initiation) on the baud clock to ascertain the position at the middle of the first data bit. It continues to read the signal at intervals of 64 cycles. After the receipt of 8 data bits it waits another 64 cycles before testing that the line level is 1 to confirm that the Stop Bit is present. If it does not find 1 it reports a framing error as already mentioned. There is some latitude in the timing of the baud clock, as stated above, but the clock must be sufficiently accurate that it neither misses a bit nor counts a bit twice.

Finally, remark that the attachment of a Start Bit and a Stop Bit to each character reduces the rate of transmission of characters. Thus, on a 1200 baud line, the rate of transmission of 8 bit characters is not 150 per second but 120 at best.

8.5 The interface

The interface which enables the processor to establish serial communication can take many forms. Reference to specific devices can produce a lot of detail without revealing clearly the main principles. Therefore, our account will concentrate on the general properties to be expected of the interface but omitting the minutiae of how any particular device makes provision for them.

The interface stands between the processor and modem or other machine that is connected. In general, the processor will use two ports to feed the interface and to receive signals from it. One of these ports will be devoted to data. The other port will be for control signals and the like.

One piece of information that the interface has to have is a statement of the conventions adopted. Hence the processor will have to tell it the line speed, the bits per character, the parity rule, the number of Stop Bits, and the position regarding XON/XOFF and SI/SO. A byte might be allotted for the processor to indicate to the interface which of the several line speeds is involved though 4 bits may be enough. Another byte might be sent to cover the other items since the bits per character (5,6,7,8) would come within 2 bits, the parity can be coped with by 2 bits, the number of Stop Bits needs only 2 bits whereas both XON/XOFF and SI/SO need a bit each. A possible layout is illustrated in Fig. 8.3. Once the processor has sent the bytes to set up the interface it does not repeat them until another group of messages is to be communicated. In other words, the interface is set up, once and for all, by the processor at the beginning of a session. The onus for obeying the protocols then passes to the interface. It receives the serial bit stream, removes any Start and Stop Bits, tests for errors, and then assembles the character in byte form for parallel transmission to the processor. Similarly, when sending, it converts a byte to a bit stream, adding Start and Stop Bits as well as seeing that the Parity rule is observed. Notice that the processor continues to handle bytes in the normal way—all the conversion from parallel to serial and vice versa is in the domain of the interface.

Bit 7	Bit 6	Bit 5 Bit 4	Bit 3 Bit 2	Bit 1 Bit 0
SI/SO	XON/XOFF	Stop Bits	Parity	Bits/char
1 = Active 0 = Off	1 = Active 0 = Off	01 = 1 Stop bit 10 = 1.5 Stop bit 11 = 2 Stop bit	00 = No parity 01 = Odd 11 = Even	00 = 5 bits 01 = 6 bits 10 = 7 bits 11 = 8 bits

FIG. 8.3 A possible parameter setter

Two further signals travel through the control port. One is despatched by the processor to activate the interface in a particular manner and the other offers input to the processor on the state of play. The first is a controlling signal from the processor and the second is an optional input for the processor to consult on the status of the interface.

The controlling signal consists of a byte which might have the structure shown in Fig. 8.4. Bit 0 is to open the transmitter in the interface when a character is to be sent; it must therefore be set to 1 before any transmission can take place. Likewise, Bit 2 is for activating the receiver; it must be set to 1 before any serial data can be received. Bit 1 is used in conjunction with a modem (if any) to inform that the interface is ready for reception.

Two new features are listed under Bits 4 and 5. When the processor has consulted the status and an abnormality has been revealed it must, after appropriate action, be able to intimate to the interface that the status can now revert to normal. The processor does this by sending a 1 on Bit 4. On receipt of this 1 the interface does two things. It first resets the status to normal. Then it returns Bit 4 to 0 so that it is ready for any future resetting message from the processor.

The BREAK signal is a long succession of 0s such as would occur if a linking cable broke. As a controller it can be used to terminate reception. Thus, if the processor places a 1 at Bit 5 the interface issues a Break signal to the transmitting device and tells it to stop sending. Such a capability may be helpful if an incoming message has not been completed so that you cannot make an exit in the regular mode.

Bits 6 and 7 have been left vacant but, in any particular device, they may have designated purposes perhaps in connection with DCD, XON/XOFF, etc.

The status byte is made up of a series of flags. A possible combination is depicted in Fig. 8.5 where the significance of a flag raised to 1 is described. Bit 0 is raised to 1 when the data buffer of the transmitter is empty and ready to accept a fresh character from the processor. As soon as the processor writes a character into the data buffer the bit is cleared to 0. Bit 1 performs a similar function for reception indicating that the receiver data buffer has a character

Bit 7	Bit 6	Bit 5	Bit 4	Bit 3	Bit 2	Bit 1	Bit 0
Free		Break	Status reset	RTS	Open receiver	DTR	Open transmitter
		1 = Send 0 = Normal	1 = Reset then return to 0 0 = Normal	1 = In use 0 = Off	1 = Activate 0 = Off	1 = Ready 0 = Off	1 = Activate 0 = Off

FIG. 8.4 Possible coding for a control byte

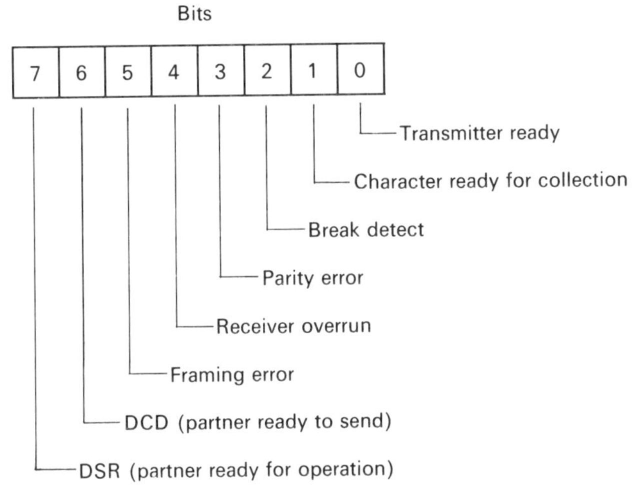

FIG. 8.5 A possible status byte

available for collection by the processor when it is raised to 1. Reading of the character by the processor clears the bit to 0.

The flag in Bit 2 is raised when it is ascertained that a Break signal has arrived from the partner in the communication. The parity error flag in Bit 3 is raised when the receiver has discovered a parity error. It is cleared by the processor putting a 1 in Bit 4 of the control byte. Likewise, Bit 5 is raised to direct attention to a framing error and is cleared via the control byte.

The flag in Bit 4 is raised when the receiver buffer has prepared a new character for transfer to the data buffer but the processor has not yet removed the previous character from the data buffer. The phenomenon is known as *receiver overrun*. Again the flag is cleared through the control byte. An extra bit (say 6) is sometimes allocated for the purpose of signalling *receiver buffer overflow*. If this happens frequently with consequent loss of characters, and

XON/XOFF is not available, either the buffer size should be increased or the speed of transmission cut. The operating systems of some microcomputers offer the user the opportunity of lowering the transmission rate by inserting time delays between blocks of data. The delays might be introduced after each character or after each carriage return or after so many bytes. As a general rule, delays are undesirable but, if characters are being lost systematically, the user can bring in delays in an empirical fashion in an effort to solve the problem.

No additional explanation of Bits 6 and 7 is necessary except to observe that if Bit 7 is raised normal communication is feasible.

Although one may gain the impression from the foregoing description that the capacity of the status byte is somewhat limited in the sense that it waits around for consultation by the processor, other arrangements may create more flexibility. Some interfaces permit direct wiring of some of the bits to the bus of the processor e.g. connecting Transmitter Ready to the INTR line. Excitation of a bit then causes an interrupt to go to the processor. In this way the attention of the processor can be attracted when circumstances warrant it without the necessity to poll that bit.

8.6 Setting up

In the prior section it has been pointed out that two bytes could be needed for determining the mode of operation of the interface (see Fig. 8.3). We shall call these two bytes SETUP1 and SETUP2. Usually, they will be supplied with default values by the operating system. If the handshaking does not match they will have to be adjusted. The normal method for modification involves the monitor and keyboard which will imply Interrupt Service Routines for such intervention. Input from the keyboard is more conveniently associated with the next chapter and so there will be no further consideration of the alteration of SETUP1 and SETUP2. They will be assumed to be located in memory (with the same names) with the correct values for the communication in mind.

Access between the processor and interface will be assumed to be through ports 10 and 20, which numbers are chosen merely to fix ideas; they have no other significance. Port 10 will be for the inward and outward movement of data. Port 20 is assigned control duties; control signals from the processor go to this port and the status byte can be received from it.

The setting up program might then take the following simple form.

```
                dseg
    control     equ     20                      ; specify port for control
    extrn       setup1:byte,setup2:byte         ; available from elsewhere
;
                cseg
    set_up:     mov     al,setup1               ; get first byte
                out     control,al              ; and send it to interface
```

```
        mov     al,setup2       ; get second byte
        out     control,al      ; and send to interface
        ...                     ; any instructions required
                                ; to inform the interface that
                                ; setting up is over
```

It is not possible to conclude this program without reference to a specific device since some are conditioned to understand that, after a certain number of instructions of particular type, all subsequent instructions will be control bytes.

8.7 Moving characters

Having set up the interface we can consider transferring characters to and from the processor. The first thing to discuss is the control byte which will be taken to have the structure of Fig. 8.4 with Bits 6 and 7 always given the value 0. For reception the transmitter will be inactive so Bit 0 becomes 0. On the other hand we must enable operation of the receiver so Bit 2 is raised to 1. We assume that no modem is taking part and make Bits 1 and 3 equal to 0. Obviously, Bit 5 is 0. With regard to Bit 4 we cannot be sure of the opening state of the flags. It is therefore a sound precaution to ensure that they are normal initially. Accordingly, Bit 4 is set to 1. Thus the control byte is 0001 0100 or 14H.

The simplest possible program is then (the status byte is taken to have the pattern of Fig. 8.5).

```
            dseg
data_port   equ     10              ; identification of route for data
cntrl_port  equ     20              ; port for control and status
in_cntrl    equ     14h             ; control signal
            cseg
simp_get:   mov     al,in_cntrl     ; get control byte
            out     cntrl_port,al   ; and send to interface
;
ready:      in      al,cntrl_port   ; bring in status byte
            and     al,2            ; check if ready for
                                    ; collection flag raised
            jz      ready           ; if not, continue polling
            in      al,data_port    ; if so, read character
            ret                     ; and depart leaving
                                    ; character in AL
```

The routine, after opening the receiver, just keeps waiting until the Ready flag is raised and then collects the character. There is no call for an instruction to clear the flag because that is done automatically on reading the character.

Notice that a different name has been attributed to the control port from in SET_UP. This is because some assemblers object to a symbol being redefined by a subsequent EQU or other directive once it has been defined; after all, there is EXTRN. However, I shall not adhere to this practice in the following and will feel free to repeat a name.

A few changes produce a program for transmitting a character located at CHAR. As far as the control byte is concerned the transmitter must be opened so that Bit 0 must be 1. The absence of reception means that Bit 2 is 0 and, since the error flags refer only to the receiver, Bit 4 can be 0. Hence, the control byte is 0000 0001 or 01H.

```
            dseg
data_port   equ    10              ; number of port for data
cntrl_port  equ    20              ; number of port for status
out_cntrl   equ    1               ; transmitter control
;
            cseg
sim_put:    mov    al,out_cntrl    ; get control byte
            out    cntrl_port,al   ; turn on transmitter
;
t_ok:       in     al,cntrl_port   ; enter status byte
            test   al,1            ; check if transmitter
                                   ; ready
            jz     t_ok            ; if not, poll again
            mov    al,char         ; if so, get character
            out    data_port,al    ; and send it
            ret                    ; back to caller
```

Neither of these programs is of much practical value. No instructions to check errors of transmission or to close the interface at the end have been included. Nevertheless, they are frameworks around which to build more sophisticated routines, as will be seen in later sections.

8.8 Error checking

As a step up the ladder, consider receiving a character and placing it in AH if there is no error. AL is to be returned as 00 in the absence of error and as 0FFH if an error was detected. At the end the interface is to be left ready for another character.

```
            dseg
data_port   equ    10              ; number of data port
cntrl_port  equ    20              ; port for status and control
in_cntrl    equ    14h             ; receiver signal
```

```
          cseg
get:      mov     al,in_cntrl       ; get control byte
          out     cntrl_port,al     ; turn on receiver
;
pull:     in      al,cntrl_port     ; bring in status byte
          test    al,2              ; see if receiver ready
          jz      pull              ; if not, poll
;
          test    al,38h            ; if so, check for errors
          jnz     err               ; if error go to ERR
          in      al,data_port      ; otherwise read character
          mov     ah,al             ; place in AH
          mov     al,0              ; put O.K. marker in AL
          ret                       ; and leave
;
err:      in      al,data_port      ; read character
          mov     al,in_cntrl       ; prepare to reset
          out     cntrl_port,al     ; receiver flags
          mov     al,0ffh           ; enter error marker
          ret                       ; and finish
```

The error checking is done by TEST which affects only flags but not AL. It performs an AND with 38H or 0011 1000 so that the Zero Flag is not raised unless an error has been recorded in one of Bits 3, 4, and 5 of the status byte. If no error is discovered, the character is read and passed to AH while an error-free marker is placed in AL. No resetting of the receiver is necessary since it has been left on and the error bits are clear.

When there is a fault the character has to be removed from the interface and the error bits must be cleared before the next character. This is done by repeating the opening instructions. Afterwards, 0FFH is put in AL to signal the occurrence of an error.

The next example deals with the reception of a sequence of characters and placing them one after the other in memory at BUFFER. The space alloted to BUFFER is 256 bytes and, when it becomes full, a Break signal is to be sent to the source. Each character is tested for errors. If none is found an acknowledgement (06H in ASCII code) is sent to the transmitter. If an error does occur a negative acknowledgement is sent and the character is not stored.

This time the interface has both transmitter and receiver in operation together which can be covered by marking Bits 0 and 2 of the control byte each 1. Keeping Bit 4 as 1 we have 0001 0101 or 15H for the control byte.

The form of the program is

```
          dseg
data_port equ     10                ; identify data port
```

```
        cntrol_prt   equ    20               ; identify control port
        io_cntrl     equ    15H              ; input/output control
        brk          equ    20h              ; break signal
        ack          equ    06h              ; ASCII acknowledgement
        nak          equ    15H              ; ASCII negative acknowledge
        buf_size     equ    256              ; size of BUFFER
        clear        equ    10h              ; signal for error flags

                     cseg
        rec:         mov    al,io_cntrl      ; get control byte
                     out    cntrl_port,al    ; turn on receiver
                                             ; and transmitter
;
                     xor    bx,bx            ; clear BX for counting
;
        bring:       in     al,cntrl_port    ; bring in status byte
                     test   al,2             ; check receiver ready
                     jz     bring            ; if not, try again
;
                     mov    ah,al            ; copy status
                     in     al,data_port     ; collect character
                     test   ah,38h           ; check status errors
                     jnz    no_char          ; if error, imperfect sending
;
                     mov    buffer[bx],al    ; otherwise, store character
                     mov    al,ack           ; and send
                     out    data_port,al     ; acknowledgement
                     inc    bx               ; add 1 to counter
                     cmp    bx,buf_size      ; see if buffer full
                     jnz    bring            ; if not, get next character
                     mov    al,brk           ; otherwise, send
                     out    data_port,al     ; Break signal
                     ret                     ; and finish
;
        no_char:     mov    al,nak           ; if character error, send
                     out    data_port,al     ; negative acknowledge
                     mov    al,clear         ; and clear
                     out    cntrl_port,al    ; the error flags
                     jmp    bring            ; go for next character
```

Several features should be remarked. The program is defective because the only way in which it terminates is when the buffer is full. If there is a systematic error in transmission or the source runs out of characters before the buffer is filled the polling process never ceases. It would be wise to include a safety net to prevent this eventuality.

A special signal CLEAR has been used to clear the status error flags in the event of an imperfect character instead of calling on IO_CNTRL. This is a precautionary measure in case the interface happens to be one that re-activation of the receiver destroys any data already in it. In such a contingency using IO_CNTRL could corrupt some of the data following a poor character perhaps to the extent that the program permanently cycled through the instructions for an imperfect character.

It may seem surprising that both IO_CNTRL and NAK are attributed the same value. They are, in fact, for different purposes and are not applied to the same port. NAK is a signal that goes out to the source through the data port whereas IO_CNTRL goes to the interface controller via the control port. Of course, one could contemplate one label instead of two. But the confusion that would cause in reading the program scarcely justifies the saving of one definition. The assembler will, after all, interpret these directives to make the correct substitutions and the assembler code will be neither longer nor shorter whichever the method of definition.

8.9 Sending

The reverse action of sending out a sequence of characters is the topic of this section. For a variation the length of the sequence, stored in consecutive bytes at ARRAY, is unspecified but its termination is indicated by the End of Text character (ASCII 03H). No acknowledgement signals will be issued by the destination and the interface is to be closed down at the end of transmission.

```
            dseg
data_port   equ     10              ; data port number
cntrl_port  equ     20              ; control port number
out_cntrl   equ      1              ; open transmitter code
close       equ      0              ; close down code
etx         equ      3              ; end of text code

            cseg
put:        mov     al,out_cntrl    ; activate the
            out     cntrl_port,al   ; transmitter
;
            lea     si,array        ; address of characters
                                    ; to SI
;
pick_up:    mov     al,[si]         ; character to AL
            out     data_port,al    ; send it
            cmp     al,etx          ; is it ETX?
            jz      last            ; if so, end of the array
            inc     si              ; if not, add 1 to SI
            jmp     pick_up         ; and get next character
```

```
;
last:       in      al,cntrl_port    ; bring in status byte
            test    al,1             ; is the transmitter empty?
            jnz     last             ; if not, poll again
            mov     al,close         ; otherwise
            out     cntrl_port,al    ; close down
            ret                      ; and finish
```

The program is a straightforward transfer of the characters, one at a time, by means of a pointer which is regularly updated until the occurrence of ETX. After the despatch of ETX the processor moves to the instructions labelled LAST and waits until the status byte confirms that the transmitter is empty. It then closes down. The success of this last part cannot be unambiguously verified because of the two buffers in the transmitter. If the sole aim of the status flag is to indicate that the transmitter data buffer is free then the close might take place before the transmitting buffer had disgorged its contents. No problem arises if the flag being 0 asserts that both buffers are empty. A solution in the first set of circumstances, if it is found that characters are being trapped in the transmitter, is to incorporate a delay in the instructions before issuing the close down.

Exercises

1. Calculate the number of ASCII characters that can be sent per second over the lines with speeds given in Section 8.3.
2. What latitude can you allow in the timing of the baud clock without its failing to perform its sampling function satisfactorily?
3. Errors affects the bits of a byte randomly and independently. If the probability of an error in a bit is 0.0001 find the probability that there is (a) no error in a byte, (b) one error in a byte, (c) two errors in a byte. Do you think parity checking would be profitable in these circumstances?
4. It is arranged that, if a 1 appears in the status byte, an interrupt is issued to the processor. Write an Interrupt Service Routine that will close the interface if a Break signal is received but ignore other interrupts.
5. If RET were altered to IRET in the program GET would a suitable Interrupt Service Routine be produced?
6. Modify GET so that, if a Break is detected, 0FH is placed in AL and the receiver turned off.
7. Alter REC so that when an end-of-file character (1AH) occurs reception ceases. As a further refinement, arrange that the Break signal is sent as soon as BUFFER is three quarters full but process any data in the receiver before closing down.
8. Often one desires to have lower case letters translated automatically to upper case. Put instructions in REC to accomplish this while ignoring

characters greater than 7FH but leaving other characters unaltered.

9. Bit 1 of the status byte is wired to INTR so that it produces an interrupt for the processor when it is 1. A sequence of incoming characters is to be placed in BUFFER. Write an Interrupt Service Routine with a similar purpose to REC. Will any registers be denied to the processor between interrupts?

10. In PUT the transmitter is not to be turned off until the characters ACK ETX have been returned by the partner in the communication. Alter the instructions to allow for this.

11. A character is not to be sent from ARRAY until an ACK is returned by the partner on receipt of the previous character. Change the program to cover this possibility. An extra check is that, if a NAK is returned signifying an impaired character, the previous character is to be sent again. Adjust your program to cope.

12. Bits 0 and 1 of the status byte are wired to produce interrupts. Repeat Exercise 11 but with an Interrupt Service Routine. Will you need a separate program to open the interface?

13. To assist in error-free transmission each character in an incoming message is repeated. If the two copies agree the character is accepted as correct and ACK is sent; if they disagree NAK is sent whereupon two further copies will arrive. Store the correct message in an array at MESSAGE. Write an Interrupt Service Routine, Bits 0 and 1 of the status byte being as in Exercise 12.

14. The time is kept in a byte at each of HOUR, MINUTE, SECOND, and a byte contains two BCD digits. Write a program to send the time as six ASCII bytes.

15. In a *parity-check code* a byte contains 5 data bits and 3 check bits. Design a system to keep the data as error-free as feasible after transmission.

9

SYSTEM CALLS

9.1 Generalities

Most microcomputers come with a *Basic Input/Output System* (BIOS) which acts as an interface between the operating system and input/output devices such as the screen and keyboard. There will also be a *Basic Disk Operating System* (BDOS) for managing disk files. The absence of an actual physical disk in the configuration does not invalidate this statement because the ROM and part of the RAM are often treated as disk devices, i.e. part of the memory is used as if it were a disk drive. In fact, the part of RAM involved as a pseudo-disk drive is commonly called the *RAM disk*.

The duty of BIOS is, as its name suggests, to provide prepared programs which look after input and output. There are many mundane features associated with these programs because each output device and each input device is connected to its own port on the microprocessor. So, a typical transfer of a character has two elements—a placing of the character at a port and a shift of the character from the port to its destination. For input, the keyboard, for example, puts the character at its port and then the processor reads it while, in output, the processor places the character at the port ready for action by the output device. The operation breaks down into subsidiary procedures. For instance, in sending a character to the screen it must be checked that the screen is ready to accept a character, any necessary scrolling must be performed, the character has to be located in the right register and then moved to the relevant port. These jobs are discharged by interrupt service routines (see Chapter 7), access being gained through the vector table for interrupts. Generally, these routines are a permanent part of the machine in order that software can be loaded and messages displayed before and during loading.

After the computer has been switched on and the operating system loaded, BDOS becomes available. Its functions are also controlled by interrupt vectors. Broadly speaking, the purpose of BDOS is to provide facilities that are not in BIOS or to combine BIOS routines into a format more convenient for the user. Any input or output required by an assembly program will be channelled through BDOS and BIOS (unless you are writing your own interrupt service routines). It is usually considered preferable to have BDOS as the intermediary and we shall follow that procedure in this chapter. In fact, we shall invoke only one of the interrupts of BDOS but it comes with many branches.

Assembly programming in the environment of existing BIOS and BDOS is consequently fairly common. This chapter considers some of the applications in this situation. What facilities you actually get in BIOS and BDOS depends upon the supplier. Indeed, they are likely to contain several of the programs, probably in a more comprehensive guise, that have been examined in earlier chapters.

In view of the varied extent of the software that is provided, our discussion will be limited to a few basic features which you would expect of any operating system. Most operating systems will offer a good deal more though the implementation of some features may need a deeper knowledge of the system that we wish to assume.

The fundamental mechanism consists of preparing the registers, putting the code number of the function in AH and then calling upon the system to perform the appropriate action. The function numbers and associated registers tend to vary slowly, if at all, with versions of the software but it is sensible to check that the values in your version are compatible with ours. The most likely deviation is that CL is employed for the location of the function number rather than AH. Also you may find that INT 224 takes the place of our INT 21H. At any rate if you find that our code does not work for you it is worth trying the switch of AH to CL and/or INT 21H to INT 224.

Another general point concerns the availability of CONTROL-C (03H in ASCII) or ^C; some keyboards have different but equivalent labels. A user may type CONTROL-C to make an exit from input through a keyboard or output to a display monitor or printer. If a function grants this facility control generally returns to where it was when the original call to the operating system was made and your program ends. In addition, a carriage return and line feed may be sent to the display so that resumption of activity is at the beginning of the next line.

9.2 Input/output

The descriptions of the functions which follow will be in module form so that they can appear in programs by means of a CALL. Later on, fuller programs will illustrate how the modules can be combined. At the beginning of each module there is an account of the properties and registers involved.

A primitive requirement is to display a character on the screen, so this is our first example.

```
; ******************************
; 02H—Display character
; Entry: AH=02H, DL=character
; CONTROL-C is available
; ******************************
;
display_char:  mov   dl,character   ; put character in DL
```

```
        mov   ah,02h        ; function code in AH
        int   21h           ; use operating
        ret                 ; system and leave
```

In this fragment the item to be displayed is assumed to be located at CHARACTER. A different location can be used by altering the first instruction suitably. Alternatively, the main program may be responsible for putting the character in DL, in which case the first instruction can be omitted except for the label which must be transferred to the second instruction. The instruction INT provides a software interrupt; the number following indicates which of 256 possible interrupts has been selected. The number 21H corresponds to function requests. Indeed, the more function requests that can be called on the less need there is for other interrupts.

To send the output to the printer only the function number has to be changed:

```
; ******************************
; 05H—Print character
; Entry: AH=05H, DL=character
; CONTROL-C is available
; ******************************
;
  print_char:  mov   dl,character  ; put character in DL
               mov   ah,05h        ; function code in AH
               int   21h
               ret
```

Another function allows the display of a message string such as ‘Disk error’. This has first to be declared in memory with a terminal ‘$’ to indicate a string for display e.g.

```
string   db   'Disk error', 0dh,0ah, '$'
```

The 0DH and 0AH when sent to the display will cause a carriage return and line feed. If you do not want either of these events to occur then you should omit it from the declaration. The string is displayed but not the $.

```
; ******************************************
; 09H—Display string
; Entry: AH=09H, DX=offset of string ending in '$'
; ******************************************
;
  display_str:  mov   dx,offset string  ; offset to DX
                mov   ah,09H            ; function to AH
                int   21H
                ret
```

The new directive OFFSET tells the assembler to insert the offset of the address of the location of STRING. Since STRING is declared as data the offset will be from the segment address in DS.

For input from the keyboard we have

```
; ***********************************************************
; 08H—Read keyboard
; Entry: AH=08H
; Exit: AL=character from keyboard
; Function waits for character to be typed. CONTROL-C available.
; ***********************************************************
;
read_kbd:   mov   ah,08h      ; function to AH
            int   21h
            ret
```

The function does not *echo* (i.e. copy) the character to the display when it is placed in AL. If you want the echo then function 01H should be used instead.

```
; *********************************************
; 01H—Read keyboard and echo
; The same as 08H but echos character to display
; *********************************************
```

As an example consider entering characters from the keyboard and changing any in lower case to upper case before displaying them.

```
begin:      call  read_kbd      ; wait for character
            cmp   al,'a'        ; if less than a not
            jb    uppercase     ; lower case so display
            cmp   al,'z'        ; if above z, not
            ja    uppercase     ; lower case
            sub   al,20h        ; otherwise, change to ASCII
                                ; uppercase
uppercase:  mov   character,al  ; prepare for display
            call  display_char  ; and show it
            jmp   begin         ; next character
```

Here we have assumed that the function modules are present in another part of the program. If they were not then the instructions would be substituted for the calls. A direct transfer from AL to DL for the display would then be feasible instead of having to go through CHARACTER.

Another illustration is obtained from a request for a password of up to 8 characters; the end is signalled by a carriage return.

```
string      db      'Password?',0dh,0ah,'$'
                                                ; prompt for query
password    rs      8                           ; space to
                                                ; hold input
begin:      call    display_str                 ; show prompt
            mov     cx,8                        ; set maximum
            xor     bx,bx                       ; prepare BX to
                                                ; point to storage
get_pwd:    call    read_kbd_and_echo           ; use function 01H
            cmp     al,0dh                      ; is it carriage
                                                ; return
            jz      exit                        ; if so, finished
            mov     password[bx],al             ; otherwise store
            inc     bx                          ; increment BX
            loop    get_pwd                     ; go for next
                                                ; character, ignoring
                                                ; more than 8
exit:       ret
```

The foregoing example reads the characters from the keyboard one by one. It is possible to enter a string of characters by means of function 0AH.

```
; ************************************************************
; 0AH—buffered input
; Entry: AH=0AH, DX=offset of buffer
; Exit: String in buffer
; Function waits for characters to be typed until carriage return is pressed.
  Characters are
; echoed to the display and
; string can be edited while being entered. CONTROL-C is
; available.
; ************************************************************
```

Amplification of how the buffer is formed is necessary. The first byte of the buffer states the maximum amount of storage available to the string. Since the largest number that can be placed in a byte is 255 the space for storage cannot exceed 255 bytes. In fact, the terminating carriage return is stored so that the input string is limited to 254 characters at most. The second byte of the buffer records the actual number of characters entered, excluding the carriage return. From the third byte onwards the incoming string is stored. Thus the buffer must consist of 3 bytes more than the maximum length of the input.

To avoid wasteful allocation of memory the first byte of the buffer should be set as low as feasible but, in the case of doubt, should be made 0FFH. If an attempt to insert more characters than specified by the buffer is made the additional characters are ignored and the bell or beep is sounded so that, effectively, the system is waiting until you press carriage return.

Notice that the input can be edited via back spacing, etc., if desired.

The following program accepts up to 30 characters and then displays the string that has been entered

```
        buffer    db      31                   ; storage for 30 chars+CR
        inchars   rs      1                    ; space for number of chars
        string    rs      31                   ; space for input
;
        buf_in:   mov     dx,offset buffer     ; set up DX
                  mov     ah,0AH               ; function to AH
                  int     21h
                  xor     bx,bx                ; prepare BX for index
                  mov     bl,inchars           ; insert length of string
                  mov     string[bx],'$'       ; add string terminator
                  call    display_str          ; invoke function 09H
```

Observe that there are 33 bytes in the buffer in accordance with the calculation earlier. The first byte is given the value of 31, the maximum size of storage required for 30 characters plus carriage return. The second byte has been reserved for the count and named INCHARS. The third and following 30 bytes have been set aside for the input; this group has been named STRING. These two names have been included for clarity and to improve the instructions. We could have omitted them and referred to the locations as BUFFER + 1 and BUFFER + 2, though then we could not have called on function 09H in such a simple fashion as we have.

The purpose of the three instructions involving BX is to replace the carriage return following the string by '$', as needed by 09H. The operation of 0AH ensures that the byte at INCHARS has been filled by the number of characters in the input string.

9.3 Macros

All of the system calls in the preceding section end with the same two instructions, an interrupt and a return. They are conveniently short but if the interrupt number were to be altered every one of the five function requests would have to be rewritten. Such an eventuality can be evaded by putting them in a separate module somewhere in the program. For instance, if

```
        bdosi        equ    21h
```

appears in the data declarations, and the instructions

```
        bdos:        int    bdosi
                     ret
```

are somewhere in the code we could modify DISPLAY_CHAR to

```
display_char: mov dl, character
              mov al,02h
              call bdos
```

By following this pattern throughout the function requests only one directive, the definition of BDOSI, has to be adjusted to apply a new interrupt to all.

This is one way of introducing flexibility into programs and making them less prone to mistakes if you have to re-identify an instruction which exists in many places. Another source of flexibility when portions of program keep reappearing with slight variations is the *macro.*

A macro is a program containing parameters which are at your disposal or, rather, parameters which you can specify in advance of using the macro—you are not permitted to leave any of the parameters undefined. Thus, a macro is helpful when the same sequence of instructions is present in several places with variations in only one or two elements. Note that a macro is not a module or sub-routine because it is not initiated by a CALL. It is more accurate to regard a macro as an assembler directive running over several lines of code because, during assembly, all the lines of the macro are translated with the parameters replaced by your specification.

DISPLAY_CHAR could be set up as a macro by

```
display_char  macro  character
              mov    dl,character
              mov    ah,02h
              int    21h
              endm
```

Observe that the name appears on the left of MACRO and the parameter on the right; if there were more than one parameter they would be separated by commas. The final ENDM designates the finish of the macro.

The presence of a parameter is not obligatory e.g.

```
read_kbd    macro
            mov      ah,08h
            int      21h
            endm
```

With these two macros the instructions for converting lower case to upper case in the foregoing section can be formulated as (comments being omitted)

```
begin:      read_kbd
            cmp    al,'a'
            jb     uppercase
            cmp    al,'z'
            ja     uppercase
```

```
                    sub     al,20h
uppercase:          display_char al
                    jmp     begin
```

Observe the positions of the names of the macros in these instructions. In accordance with the notion that they are extended directives they are in the column for the mnemonic of an instruction. Any accompanying parameter goes into the column for operands. The freedom with parameters is exemplified in UPPERCASE where the character sought is in AL; the assembler will see that the operand CHARACTER is replaced by AL throughout the macro.

As a further example of a macro we give one for obtaining the address pointing to an interrupt routine.

```
; ********************************************
; 35H—Get interrupt vector
; Entry: AH=35H, AL=Interrupt vector number
; Exit: ES:BX is address of interrupt handler
; in notation of Section 2.2
; ********************************************
;
  get_vector  macro  interrupt
              mov    al,interrupt
              mov    ah,35h
              int    21h
              endm
```

The practice of having the same name in these two sections e.g. READ_KBD for both a macro and a set of instructions is absolutely unsound. Different labels are essential to distinguish between sets of instructions which are called and directives which are not. They have been kept the same here to assist with cross-referencing between the ideas.

Not all assemblers support macros; different assemblers accommodate different directives. That is why the function requests in the preceding section have not been stated as macros. Instead they have been given in forms that are applicable whether or not the assembler can cope with macros. As you can see it is quite easy to convert them to macros. Therefore, future function requests will not be listed as macros; it will be for you to make the conversion if you wish but remember to keep their names distinct from any labels which you use for calling.

9.4 Files

Moving files to and from disk entails more complicated manoeuvres than the simple operations of Section 9.2. Advances in software enable simplification by concealing some awkward aspects. However, not all operating systems contain these advances and there is some virtue in relating what goes on to analogues in high-level languages, like opening and closing files. The question of maintaining a directory of files on disk will not be addressed here. Nor will the *File Allocation Table* (FAT), which converts clusters of a file to disk sectors, be touched on since it is mainly of interest when writing device drivers. Our topic is the more mundane business of handling an individual file.

A place in memory has to be set aside for identifying the file for subsequent reference. This is done by the operating system when an external command is typed or a program is executed. The lowest available free memory is then allocated and constitutes the *Program Segment*. The first 256 bytes of the Program Section are devoted to the purposes of the operating system and do not contain the instructions of your program; this block of 256 bytes is therefore called the *Program Segment Prefix*. Within the Program Segment Prefix two sections are reserved for file information which can be drawn upon during file operations. Each of these sections is known as a *File Control Block* (FCB). They are positioned at offsets 5CH and 6CH from the start of the Program Segment.

The aim of a File Control Block is to sustain an up-to-date account of the file's name, size, date, record length, etc. In our context the most relevant part is the first 12 bytes, which has three subdivisions. The first byte holds the drive number (A = 1, B = 2, . . .) and specifies the disk drive; an Open File command sets it to the number of the default drive. The next 8 bytes contain the filename, padded with blanks if necessary. The remaining 3 bytes are for the extension; again there is padding with blanks if necessary and the 3 characters are permitted to be all blank when there is no extension.

An FCB may be either *opened* or *unopened*. An unopened FCB is restricted to the drive specifier and filename; this is located at offset 5CH (or 6CH) in the Program Segment Prefix. When the FCB is opened, further information on the file size, record size (normally 128 bytes), date and time when the file was created or last updated, etc. is inserted at a suitable place in the Program Segment Prefix.

In this connection it should be noted that when a command is entered the first parameter goes to the FCB at 5CH and the second (if any) to the FCB at 6CH, both being unopened. If either parameter is a pathname, the corresponding FCB contains only the valid drive number; the filename will not be valid.

The system call to open a file is

```
; *********************************************
; 0FH—Open file
; Entry: AH=0FH, DX=offset of unopened FCB
; Exit: AL=0 File found and FCB filled
; AL=0FFH File not found
; *********************************************
```

```
;
  open:       mov    dx,offset fcb
              mov    ah,0FH
              call   bdos
```

where we have used BDOS as described in Section 9.3.

To close a file issue 10H. However, if this coincides with ending a program it is better to employ 4CH.

```
; *******************************************
; 10H—Close file
; Entry: AH=10H, DX=offset of opened FCB
; Exit: AL=0 File closed
; AL=0FFH File not found
; *****************************************
;
  close:      mov    dx,offset fcb
              mov    ah,10h
              call   bdos
```

It may be that you desire an FCB independent of those of the operating system. Such an FCB can be generated by means of a data declaration. Suppose that the title of the file is ALPHABET.TXT. Then the pertinent declaration is (if the FCB is named FCB1)

```
              fcb1          db    0, 'ALPHABETTXT'
                            rs    26
```

This sets the first byte to 0, intimating that the default drive is relevant. The next eleven bytes are set to the string giving the filename and extension; notice that the full stop antecedent to the extension is omitted. The succeeding 26 bytes are reserved without any initial values. They are spaces to be filled in when the file is opened.

If the file had been ALPHA.TX, blanks would have to be inserted to complete the eleven bytes as in

```
              fcb2          db    0, 'ALPHA    TX '
                            rs    26
```

When ALPHABET.TXT exists the FCB can be opened by means of the function 0FH. For a new file another system call is put in train.

```
; ***********************************************************
; 16H—Create file
; Entry:  AH=16H, DX=offset of unopened FCB
; Exit:  AL=0 New file opened (any existing file with the same name
;        has been erased)
;        AL=0FFH No space for another directory entry
; ***********************************************************
```

```
;
create:     mov   offset fcb1
            mov   ah,16h
            call  bdos
```

Observe that CREATE opens automatically the file so that there is no need to issue the system call 0FH as well.

A file is viewed as composed of records all of the same size. At the end of a file a record may not be entirely full so that a partial record is formed. In some operations a partial record may be expanded to record size by padding with zeros. As already stated, the operating system sets a default value of 128 bytes on the record size. To have a different size you have to know that the record size is offset 0EH from the start of the FCB. Suppose the file is to consist of a sequence of disk numbers plus filenames. The appropriate record size would then be 12 bytes. The instruction

```
mov   fcb+0eh,12
```

allows you to work with this record size. Alternatively, you may prefer a directive plus an instruction as in

```
record_size   equ   0eh
              mov   fcb+record_size,12
```

Again, if a particular record within a file is desired, this can be pointed to from offset 21H in the FCB. For example, access to the third record is achieved by

```
mov   fcb+21h,2
```

remembering that the first record is numbered 0.

9.5 Disk transfers

The fundamental element in moving files to and from a disk is the record, i.e. a read or write instruction transfers one record. Consequently, to move a whole file you must have as many read or write instructions as there are records in the file. Such repetition is a natural candidate for cycling with the end-of-file marker serving as a sentry to tell you when to stop.

The normal means for accomplishing a transfer is to establish a buffer which will accommodate a record. One obvious consequence is that the magnitude of the buffer should be at least as large as the size of the record. It is not uncommon to make the buffer 128 bytes, the default value of the record size. Of course, a smaller buffer is permissible when you are confident that the record will fit into it.

The operating system has to be informed where the buffer is in order to arrange the transfer of data between it and the disk. The information is conveyed by supplying the address of the buffer, called the *Disk Transfer Address* (DTA). A system call is provided for this purpose and, in case you forget to set the Disk Transfer Address, the system has a default location in offset 80H of the Program Segment Prefix to prevent hang-ups.

```
; ************************************************************
; 1AH—Set Disk Transfer Address
; Entry: AH=1AH, DX=DTA
; Overflow into another segment is not allowed
; ************************************************************
;
set_dta:    mov     dx,offset buffer
            mov     ah,1ah
            call    bdos
```

Once the DTA has been set, a file can be read by the following system call.

```
; ************************************************************
; 14H—Sequential read
; Entry: AH=14H, DX=offset of opened FCB
; Exit:  AL=0 Record read into buffer
;        AL=1 End-of-file, record empty
;        AL=2 No read because buffer too small
;        AL=3 End of file, partial record padded with
;             zeros to record size
; ************************************************************
;
seq_read:   mov     dx,offset fcb
            mov     ah,14h
            call    bdos
```

The function to write to a disk is set out below

```
; ************************************************************
; 15H—Sequential write
; Entry: AH=15H, DX=offset of opened FCB
; Exit:  AL=0 Record transferred
;        AL=1 No write because disk full
;        AL=2 No write because buffer not large enough
; ************************************************************
;
seq_write:  move    dx,offset fcb
            mov     ah,15h
            call    bdos
```

As an example, we formulate a program to look at the first byte of the file

NUMB.BAS on disk B; if it is 0FFH the message *Not saved in ASCII* is to be displayed.

```
string    db    'Not saved in ASCII', 0dh,0ah,'$'
                                          ; declare message
fcb       db    2, 'NUMB        BAS'     ; prepare FCB with file name
          rs    26                        ; and spare space
buffer    rs    80h                       ; make buffer 128 bytes
;
begin:    call  set_dta                   ; set DTA of buffer
          call  open                      ; open FCB
          call  seq_read                  ; read first record
          cmp   buffer, 0ffh              ; is first byte 0FFH?
          jnz   over                      ; if not, finished
          call  display_str               ; if so, display message
over:     call  close                     ; close FCB
```

As you can see the program consists essentially of a series of system calls to open the file, read from it and then close the file after displaying a message when appropriate. You might find it interesting to express the same thing in macros.

A word of warning is in order concerning closing a file if a disk has been changed while the file was open. The system may then not be able to make the correct adjustments to the disk directory. In that case 0FFH will be returned in AL as described in Section 9.4 but it may not be possible to recover the alterations which have been made to the file.

9.6 A sample program

Most of the programs that have been written hitherto have really been fragments. In this section there is a sample in which the full picture is painted. Any system calls will have to be set out in their entirety so that the program is complete in itself. However, before getting to that point, we give the system call for an exit from a program.

```
; ***********************************************************
; 4CH—End process
; Entry: AH=4CH
; All open FCBs are closed, the current process is ended
; and control returns to the invoking program
; ***********************************************************
;
  end_process:  mov    ah,4ch
                call   bdos
```

The program to be constructed is to display the contents of a file (saved in ASCII) on the screen. Hence it is similar to the command TYPE that one encounters in high-level languages. Not surprisingly, if the file is not in ASCII format some weird and peculiar effects will be produced on the screen. The display will attempt to convert all hexadecimal numbers into printable ASCII characters. Files not in ASCII format will not produce anything recognizable. In particular, BASIC programs stored in compressed binary format will occasion strange results on the monitor unless specially saved in ASCII.

Because of the similarity of TYPE but to escape confusion with that command our title will be EPYT with the letters in reverse order. In the program we shall attempt to use labels similar to those in earlier sections so that you can see how system calls are inserted in this self-contained program.

By means of a screen editor or wordprocessor the file EPYT.A86 is created, the extension A86 signifying that it is destined for the assembler of the 8086. (The extension .ASM is appropriate for some assemblers.)

```
; ***********************************************************
; epyt.a86    —to display the contents of an ASCII
;              file on the monitor
; ***********************************************************
;
              dseg
              org    100h
bdosi         equ    21h     ; interrupt number for system calls
bsize         equ    80h     ; value for size of buffer
displayc      equ    2       ; system call for display character
end_processc  equ    4ch     ; system call for end
eof           equ    1ah     ; end-of-file number
fcb           equ    5ch     ; offset of system FCB
openc         equ    0fh     ; system call for open file
print_stc     equ    9       ; system call for display string
readc         equ    14h     ; system call for read from disk
setdtac       equ    1ah     ; system call to set DTA
bptr          db     bsize   ; declare pointer to buffer size (80h)
buffer        rs     bsize   ; reserve 128 bytes for buffer
errorm        db 'No file found', 0dh,0ah, '$'
                             ; error message for display if
                             ; no file found
error2        db 'Faulty read', 0dh,0ah, '$'
                             ; message when problem in reading
                             ; disk
;
; ***********************************************************
;
```

```
          cseg
start:    mov   dx,fcb            ; prepare to open file
          call  open              ; and open it
;
cycle:    call  get_chr           ; put next character in AL
          cmp   al,eof            ; is it end-of-file marker?
          jz    done              ; if so, finished
          mov   dl,al             ; otherwise, prepare to display
                                  ; character
          call  display_chr       ; and display it
          jmp   cycle             ; go for next character
;
done:     move  ah,end_processc   ; prepare to end
          jmp   bdos              ; and finish
;
open:     mov   ah,openc          ; insert open code in AH
          call  bdos              ; and open FCB
          cmp   al,0ffh           ; was file not found?
          jz    err               ; if so, report error
          ret                     ; otherwise, back to caller
;
bdos:     int   bdosi             ; interrupt for system
          ret                     ; calls
;
err:      mov   dx,offset errorm  ; error message for no file
                                  ; address to DX
          call  print_st          ; call to display message
          jmp   done              ; and close down
;
print_st: mov   ah,print_stc      ; code to display string
          jmp   bdos              ; display it
;
;
get_chr:  cmp   bptr,bsize        ; see if buffer pointer is
                                  ; up to buffer size
          jb    get1              ; if not, go for next character
          call  fill_buff         ; otherwise, refill buffer
                                  ; and reset buffer pointer
get1:     mov   bl,bptr           ; buffer pointer in BL
          xor   bh,bh             ; make BX an index
          mov   al,buffer[bx]     ; move next character to AL
          inc   bptr              ; increase buffer pointer
          ret                     ; and return to caller
```

```
;
    fill_buf:     mov    dx,offset buffer    ; get buffer address
                  call   set_dta             ; and use it for DTA
                  mov    dx,fcb              ; set up for read
                  call   seq_read            ; from file
                  mov    bptr,0              ; put pointer to zero
                  ret                        ; and go back to caller
;
    set_dta:      mov    ah,setdtac          ; code to set DTA
                  jmp    bdos                ; set DTA
;
    seq_read:     mov    ah,readc            ; code to read a record
                  call   bdos                ; read it to buffer
                  cmp    al,0                ; was the read OK?
                  jnz    check               ; if not, check for error
                  ret                        ; if O.K. return to caller
;
    check:        cmp    al,3                ; was it a partial record?
                  jnz    err2                ; if not, error occurred
                  ret                        ; otherwise, carry on
;
    err2:         mov    dx,offset error2    ; error message address
                                             ; for faulty read in DX
                  call   print_st            ; display it
                  jmp    done                ; and close down
;
;
    display_chr:  mov    ah,displayc         ; system call to
                  jmp    bdos                ; display character
                  end
```

The overall strategy consists of opening the FCB and then a cycle in which the next character in the file is placed in AL followed by display unless it is the end of the file when the process is terminated. Each of these elements is broken down into a number of subsidiary steps.

The FCB is opened by a system call (see the instructions labelled OPEN). Immediately afterwards, a check is made that a file has been successfully found. If it has not, control is transferred to an error routine ERR which, by means of a system call to display a string, puts the message 'No file found' on the screen and then makes an exit from the program.

The process for obtaining the next character GET_CHR contains several parts because it has to arrange for the filling of the buffer. It does this through a buffer pointer BPTR which is steadily incremented so as to indicate the next character in the buffer for transfer. When it reaches 128 the last character in the buffer has been handled and it is time to refill the buffer. So GET_CHR starts by checking if BPTR is less than 128. If it is the next character is transferred by GET1, using BX as an index, and then BPTR is updated. If BPTR equals 128 (and this is true initially because of the declaration in the data segment) the buffer is filled from the file.

Filling the buffer is achieved by FILL_BUF. First the DTA is set to the address of the buffer. Then the next record on the file is read by SEQ_READ (Remark: after each read the record pointer is automatically incremented so that successive calls to SEQ_READ bring in consecutive records). Within SEQ_READ a check is made that the read has been completed satisfactorily before returning to FILL_BUF. If a failure occurs an error message 'Faulty read' is shown on the screen and the program closed down. Notice that it has been assumed that the sizes of the record and buffer match. The final action of FILL_BUF is to start BPTR off at zero so that GET1 will work systematically through the buffer.

As many codes as possible have been named in the data segment. Two advantages flow therefrom. One is that the names, selected to intimate the purpose of the function, improve the readability of the subroutines. The other is that a change of code number involves only the alteration of one directive in the data segment; the code segment can be left intact.

It may surprise you that no file is specifically mentioned in the program. The reason is that the FCB of the operating system is used. After the program has been assembled and linked there will be a file EPYT.EXE (for MS/DOS) or EPYT.CMD (for CP/M) on disk (make sure there is enough disk space for all this). It is this file which is responsible for implementing the program. Thus the command

EPYT ALPHA.TXT

in which the extension of EPYT is omitted will result in the filename ALPHA.TXT being placed in the relevant parts of the FCB at offset 5CH of the Program Segment Prefix. So this file will be picked up by the opening instruction in START and its contents will be displayed on the monitor.

The directive END has been used in its simplest form to tell the assembler that the end of the source file has been reached. It is capable of an extension that passes on a priority which indicates which instruction is to be executed first when the program is loaded by replacing it by

```
end    start
```

START being the label of the first instruction of our code segment. The address of the label goes to the linker and is installed in the run file (see Section 9.11).

In the absence of a label for the commencing instruction, execution begins at the beginning of the first code segment. For our program this coincides with the point designated by the above directive since there is only one code segment. If you do aim to combine several programs, each with its own code segment, or have more than one code segment in a program, you should put a starting label with END on one, and only one, of them. This particular code segment will then take precedence over the others and execution will commence at the specified instruction in it. This ensures that the control of multiple code segments is firmly established.

9.7 STDIN and STDOUT

Standard input (STDIN) and *standard output* (STDOUT) are contrivances to increase the power of the programmer in manipulating input and output. A separate section deals with them because they may not be available with all systems.

STDIN and STDOUT behave as if they were universal junction boxes. For example, suppose that all output is sent to one side of STDOUT. On the opposite side is a universal coupling to which may be attached a screen, file, etc. The decision as to which is to be attached does not have to be made until a command is issued. So, on the first occasion of the command, you may direct the output to the monitor and on the next to a file on disk. The ability to shift the destination of the output at will is known as *redirection*.

Likewise, STDIN offers the possibility of redirecting input so that, for instance, at the time of command you may elect to take input from the keyboard or a disk file.

The function call for STDOUT is

```
; ************************************************************
; 40H—Write to file or device
; Entry:  AH=40H, BX=01H, CX=number of bytes to be written,
;         DX=offset of source
; Exit:   If CF=0 then AX=number of bytes written
;         If CF=1 and AX=5 access has been denied
;         If CF=1 and AX=6 there is an invalid file
; Remark:   If CF=0 but AX≠CX on exit the error is
;           probably that there is inadequate space on the disk
; ************************************************************
;
  write_file:  mov   bx,01H
               mov   dx,offset source
               mov   cx,bytes
               mov   ah,40h
               call  bdos
```

The corresponding program fragment for STDIN is

```
; ***********************************************************
; 3FH—Read file or device
; Entry:  AH=3FH, BX=00H, CX=number of bytes to be read,
;         DX=offset of buffer
; Exit:   If CF=0 then AX=number of bytes read
;         If CF=1 and AX=5 access has been denied
;         If CD=1 and AX=6 there is an invalid file
; Remark:  If CF=0 on exit the number
;          of bytes in CX may not have been transferred to the buffer;
;          Some brief reasons follow (see below for more detail). One
;          possibility is that you tried to read starting at
;          the end of file when AX returns zero. Another arises
;          in reading from the keyboard when input ceases at the
;          first carriage return.
; ***********************************************************
;
read_file:  mov   bx,00h
            mov   dx,offset buffer
            mov   cx,bytes
            mov   ah,3fh
            call  bdos
```

With function 3FH, characters are accepted from the keyboard until one of three things happens: (a) a carriage return is received (as for function 0AH), (b) the number of characters entered is 1 less than the value in CX; thereafter only a carriage return will be accepted, (c) 128 characters, including the carriage return, have been read. There is therefore no point in giving CX a larger value than 128. When the accepted characters are placed in the buffer, a line feed is added after the carriage return. The buffer, accordingly, needs to be at least 1 byte more than CX, with a maximum of 129 bytes. Note that the characters entered, as well as the carriage return and line feed, are echoed to the screen.

For an example imagine that there is a sequence of strings, each terminated by a carriage return and line feed, stored in consecutive bytes at SOURCE by an earlier part of the program. The total number of bytes (including carriage returns and line feeds) in SOURCE is given in the word at TOTAL. Then STDOUT may be implemented by

```
        ...
write:  mov   bx,01h
        mov   dx,offset source
        mov   cx,total
        mov   ah,40h
        call  bdos
        jc    err3            ; if CF=1 go to an error routine
        cmp   cx,ax           ; were all bytes transferred?
        jne   err4            ; if not, another error routine
        jmp   done            ; close down
```

The error routines and any associated messages for display have not been included. They can be constructed as for EPYT.

Suppose that the program is labelled STRING. Then the command

STRING ALPHA.TXT

will convert the file ALPHA.TXT to the strings specified in the first part of the program and then display them on the screen, a fresh string starting on a new line (because of the CR and LF). In contrast the command

STRING ALPHA.TXT >BETA.STR

will send the output to the file BETA.STR on disk without any intermediate display on the screen. The symbol > tells the system to redirect the output to the designation which follows.

If the file BETA.STR does not exist already, it will be created and its presence recorded in the directory for files. When BETA.STR is already in existence it will be overwritten by the output resulting from your command.

Overwriting can sometimes be avoided by a double use of the redirection symbol e.g.

STRING ALPHA.TXT >>BETA.STR

is intended to append the output to the file BETA.STR if it is present; if it does not exist it is created.

Suppose now that instructions are added to STRING so that its input can come via STDIN. Let us rename it STRINGIN to indicate this extra facility. Then the command

STRINGIN

will accept input from the keyboard and, after processing, display the output on the monitor. To send the output to a file instead of the screen the pertinent command is

STRINGIN >BETA.STR

For input from a file and output on the monitor issue the command

STRINGIN <ALPHA.TXT

The symbol < informs the system that input is to be picked up from the source immediately after the sign. The command

STRINGIN <ALPHA.TXT >BETA.STR

takes the input from ALPHA.TXT and redirects the output to BETA.STR.

9.8 File handles

Some systems offer the facility of *file handles* whereby the maintenance of a file control block can be avoided. Basically, a handle is a 16-bit numerical code for the specification of a file. Once a particular file has been allocated its code number all subsequent reference to it is through this handle. There is no necessity for you to know what the number of the handle is; that can be safely left to the internal working of the system.

An operation with a file via a handle can be broken down into three fundamental parts: (a) create a handle if one does not exist or open it if it has already been created, (b) make your adjustments to the file, (c) close the handle. Shortly, system calls will be described which effect these operations.

Devices can also have handles attached, but five codes are specifically reserved for input and output. These five special handles do not have to be opened before you use them, or closed afterwards. They are

0000h	Input device, e.g. keyboard
0001h	Output device such as monitor
0002h	Error output to monitor
0003h	Auxiliary communications device
0004h	Printer.

The codes 0000H and 0001H serve as STDIN and STDOUT respectively; consequently redirection is available with them (see Section 9.7).

The system call to create a handle is

```
; ***********************************************************
; 3CH—Create handle
; Entry:  AH=3CH, CX=file attribute, DX=offset of pathname
; Exit:  If CF=0, AX=allocated handle
;        If CF=1 and AX=3 the path has not been found
;        If CF=1 and AX=4 no handle is spare
;        If CF=1 and AX=5 access denied, perhaps because
;        the directory is full
; ***********************************************************
  create-handle:  mov   dx,offset pathname
                  mov   cx,0
                  mov   ah,3ch
                  call  bdos
```

The pathname in DX must be an ASCIZ string, i.e. a string terminated by a zero byte, which conveys full information about the file for which a handle is desired. The attribute in CX can be selected to force the file to be normal, read-only, hidden, and so on. The code 0 chosen above makes the file a standard read-write one. After successful completion of the system call the file has the attribute picked and has been opened ready for reading and writing. If the file was already in existence, the action of 3CH is to truncate its length to 0 so that it can be rewritten.

Unless you are convinced that you want to rewrite the whole of an existing file it is better to open the handle instead of creating it. The funtion for opening a handle is

```
; *********************************************************
; 3DH—Open handle
; Entry:  AH=3DH, DX=offset of pathname, AL=0 (read)
;         or 1 (write) or 2 (read and write)
; Exit:   If CF=0, AX contains handle
;         If CF=1 and AX=2 the file has not been found
;         If CF=1 and AX=4 no spare handle
;         If CF=1 and AX=5 access denied, perhaps because
;         you are trying to write on a read-only file
;         If CF=1 and AX=12, AL has not been set to one
;         of the three permitted values
; *********************************************************
;
  open_handle; mov   dx,offset pathname
               mov   al,access
               mov   ah,3dh
               call  bdos
```

Here DX must contain the offset of a pathname which is an ASCIZ string, as for function 3CH. ACCESS stands for one of the three codes 0, 1, 2, depending on the operations you intend to carry out on the file.

Once the handle has been successfully opened the question arises as to where operations commence on the file. For flexibility a *read/write pointer* is associated with an open file; it always points to the byte of the file that is to be tackled next. So when a file is opened the function 3DH aims it at the first byte of the file. As the bytes of the file are processed the pointer is automatically updated. As will be seen later there is a system call which allows you to move the pointer to a position of your own choosing.

An editor such as EDLIN has the facility that, given a filename, it opens the file if it exists and otherwise creates it. Here is how you might do the same with assembly language for the file TEXT.ASC on the disk in drive B.

```
file      db    'b:text.asc',0   ; ASCIZ string for filename
handle    rw    1                ; storage for handle
          mov   dx,offset file
          mov   al,2             ; to open as read/write
          mov   ah,3dh           ; function to open
          call  bdos
          jnc   keep_handle      ; if no error save handle
          cmp   ax,2             ; check if error no file
          jne   find_err         ; if not, go to error routine
          mov   cx,0             ; if so, create new file
```

```
                mov     ah,3ch          ; system call
                call    bdos
                jc      find_err        ; if CF=1 go to error routine
keep_handle:    mov     handle,ax       ; save successful handle
exit:           ret
```

The error routine FIND_ERR is not included. It would consist of checking which error code was in AX and then displaying an appropriate message before going to EXIT. Thus no handle would be saved unless either the creation or the opening had been successful.

For writing to and reading from a file the functions 40H and 3FH of Section 9.7 can be employed. The new feature is that BX must contain the handle of the file instead of the values laid down in Section 9.7 but that is the only change to the specification.

As for closing a file, the function 4CH of Section 9.6 will suffice if the process is being ended. Otherwise, use 3EH:

```
; ***********************************************************
; 3EH—Close handle
; Entry:    AH=3EH, BX=handle
; Exit:     If CF=0 no error
;           If CF=1 and AX=6, handle is not open or is
;           invalid
; ***********************************************************
  close_handle: mov   bx,handle
                mov   ah,3eh
                call  bdos
```

The following program opens the file TEXT.ASC on the disk in drive A, displays it on the monitor and then closes it.

```
  filename      db    'a: \text.asc',0     ; ASCIZ string
  buffer        rb    129                  ; for display
  handle        rw    1
;
  begin:        mov   dx,offset filename   ; prepare to open file
                mov   al,2
                mov   ah,3dh
                call  bdos
                jc    find_err             ; check for error in opening
                mov   handle,ax            ; save handle
  read_file:    mov   bx,handle            ; prepare to read 128 bytes
                mov   dx,offset buffer
                mov   cx,128
                mov   ah,3fh
                call  bdos
```

```
          jc      find_err          ; check for error on read
          cmp     ax,0              ; end of file reached?
          je      done              ; if so, close
          mov     bx,ax             ; bytes read to BX
          mov     buffer[bx],'$'    ; make string in buffer
          mov     dx,offset buffer
          mov     ah,09h            ; display string
          call    bdos
          jmp     read_file         ; read some more file
done:     mov     bx,handle         ; close handle
          mov     ah,3eh
          call    bdos
          jc      find_err          ; check for error on closing
```

Again the relevant error routines have been omitted, and have been assumed to be grouped together under the single heading FIND_ERR though there is no obligation to do so. You will find it of interest to compare this program with the corresponding one in Section 9.6 using a file control block but you should remember that the one in Section 9.6 is set out in much fuller detail.

9.9 Calls from BASIC

On occasion you will wish to take advantage of a module in assembly language while running a BASIC program. There are several methods for accomplishing this. We shall describe only the one that we have found the most straightforward. It places the assembly module permanently in memory before BASIC is loaded and so can be invoked at any convenient stage in the BASIC programme. Also, since it is separate from BASIC's space no special measures have to be taken to protect it from BASIC's operations.

The fundamental problem that has to be attacked is telling BASIC where to find the module since it may not always be loaded in the same spot. So the address of the module has to be lodged in a specific location known to BASIC. Your manual may contain the information that a certain block of memory has been reserved for the user. If so, you can select 4 bytes of it to hold the vector pointing to the assembly module. If you do not know whether any space has been reserved or where it is, probably the safest course of action is to pick a slot in the vector table for interrupts.

The vector table has 256 entries, far more than the system will have active. To prevent loss of program if an inactive interrupt is invoked inadvertently the quiescent entries will all point to an instruction which returns the program whence it came. In other words, many of the entries contain the same address and, when you examine its content, you will find that it consists of the single instruction IRET i.e. the first byte is 0CFH. Typically, the vector table runs from 0000:0000 to 0000:03FF; you can cross-check by comparing the addresses provided by system call 35H, recalling that the first word of an entry in the vector table will be the offset of the address and the second word the segment. Having confirmed that the majority of addresses are the same and point to IRET you can pick one of them well away from the system interrupts, which are usually bunched towards the beginning or end of the table. Let us fix on interrupt 80H which will occupy 0000:0200 to 0000:0203 inclusive. Then this is where the address of the assembly routine is inserted and where BASIC looks when it calls on the module.

The system call which places the address in the vector table is 25H.

```
; ***************************************************************
; 25H—set vector address
; Entry:    AH=25H, AL=interrupt vector number,
;           DS=segment of routine, DX=offset
; Exit:     There are no error messages
; ***************************************************************
;
set_vector:     push    ds              ; save DS temporarily
                mov     ax,seg subrt
                mov     ds,ax
                mov     dx,offset subrt
                mov     al,interrupt
                mov     ah,25h
                call    bdos
                pop     ds              ; recover DS
```

The directive SEG extracts the segment of the subroutine just as OFFSET selects its offset. Notice that, since this function involves DS, we have to preserve it at the beginning and return it to its original value at the end.

The module is to stay in the memory and so the system has to be informed how much space to allocate. This is done very conveniently by the read/write pointer. The technique enables the program to calculate its own length so that later adjustments which affect its length are automatically taken care of. The next system call permits the user to move the read/write pointer any number of bytes forwards or backwards.

```
; ***************************************************************
; 42H—move read/write pointer
; Entry:    AH=42H, BX=handle, CX,DX=distance to move in
;           bytes (a 32-bit signed number with CX the high-order
;           word), AL=0 (move distance from beginning of file)
;           or 1 (move distance from current pointer location)
;           or 2 (move distance from end of file)
; Exit:     If CF=0, DX,AX gives new position of read/write
```

```
;           pointer as 32-bit integer with DX high-order word
;           If CF=1 and AX=1, AL was incorrectly set
;           If CF=1 and AX=6, the handle was not open
; ************************************************************
;
move_ptr:   mov     bx,handle
            mov     cx,high_word
            mov     dx,low_word
            mov     al,mode
            mov     ah,42h
            call    bdos
```

The function moves the read/write pointer from a reference point, determined by AL, the distance specified in CX,DX. The movement is forward if the distance is positive and backward when the number in CX,DX is negative. Unless you plan to move the pointer outside the range ±32 767 bytes you can put the desired shift in DX and then make CX zero if DX is positive or 0FFFFH if DX is negative.

Two interesting cases arise when both CX and DX are zero. With AL = 1 the system call will leavc in DX,AX the value of the current position of the read/write pointer. With AL = 2, on the other hand, the value in DX,AX will be the length of the file. At the same time the pointer will be ready for you to add extra material at the end of the file should you so wish.

One further system call is for planting a program in as a permanent resident on termination of loading because 4CH ends a process by clearing everything out.

```
; ************************************************************
; 31H—Terminate but leave resident
; Entry:    AH=31H, DX=memory required in paragraphs (1 para-
;           graph=16 bytes)
;
; Exit:     There are no error messages
; ************************************************************
```

The value in DX specifies the amount of memory in paragraphs that is necessary to house the program. Since the length of a paragraph is 16 bytes programs which are longer than 64 K bytes can be placed in residence.

As an example we will consider keeping a copy of the display in the memory for later use. The location of the video buffer which holds the display varies from machine to machine. We will assume that it starts from F000:0000. For a black and white monitor the buffer will occupy some 4K of memory; for colour at least 16K will be filled. We will suppose that it is a black and white screen with 800H words to be copied.

The program consists of essentially two parts—a setting up module and the subroutine to be called from BASIC. The setting up portion is responsible for putting the address of the subroutine where BASIC can find it and leaving the program resident in memory. The subroutine carries out the copying of the display when called on by BASIC. To shorten the length of the presentation, numbers, rather than names, have been used in the program but it is better practice to follow the pattern of Section 9.6. The position of interrupt 80H in the vector table is where BASIC searches for the address of the subroutine.

```
            cseg
start:      push  ds               ; save DS temporarily
            mov   dx,offset subrt  ; get offset of subroutine
            mov   ax,seg subrt     ; move segment of subroutine
            mov   ds,ax            ; to DS
            mov   al,80h           ; insert interrupt number
            mov   ah,25h           ; transfer subroutine address
            int   21h              ; to vector table
            pop   ds               ; recover DS
;
            mov   dx,offset pgm    ; open this file
            xor   al,al            ; for reading
            mov   ah,3dh
            int   21h
            jc    over             ; check for error
            mov   handle,ax        ; save handle
;
            mov   bx,handle        ; prepare to find
            xor   cx,cx            ; length of file
            xor   dx,dx            ; by read/write pointer
            mov   al,2
            mov   ah,42h
            int   21h
            jc    over             ; check for error
            mov   file_len,ax      ; save length of file
;
            mov   bx,handle        ; close the handle
            move  ah,3eh
            int   21h
            jc    over             ; check for error
;
            mov   dx,file_len      ; make program resident
            mov   cl,4             ; convert length to
            shr   dx,cl            ; paragraphs
            add   dx,12h           ; safety margin
            mov   ah,31h
```

```
         int     21h             ; leave program
over:    jmp     close           ; go to error display
;
subrt:   push    bp              ; save BP on stack: BASIC entry
         mov     bp,sp           ; stack pointer to link register
         sub     sp,4            ; 4 bytes on stack for local
                                 ; variables
         mov     ax,ds
         mov     [bp]-2,ax       ; save DS
         mov     ax,subdat       ; point DS
         mov     ds,ax           ; to SUBDAT
         mov     bx,es
         mov     [bp]-4,bx       ; save ES
         mov     bx,0f000h
         mov     es,bx           ; ES points to video buffer
         mov     cx,800h         ; amount to be copied
         xor     bx,bx           ; clear BX for counting
round:   mov     ax,es:[bx]      ; copy word of buffer
         mov     mem[bx],ax      ; and store it
         add     bx,2            ; update BX
         loop    round           ; repeat copying
;
         mov     bx,[bp]-4
         mov     es,bx           ; recover ES
         mov     ax,[bp]-2
         mov     ds,ax           ; recover DS
         mov     sp,bp           ; link to stack pointer
         pop     bp              ; restore BP
         retf                    ; back to BASIC
;
close:   mov     dx,offset string
         mov     ah,9h           ; display error string
         int     21h
         mov     ah,4ch
         int     21h             ; end process
subdat   dseg    para
pgm      db      'video.exe',0
handle   rw      1
file_len rw      1
string   db      'System call has failed','$'
mem      rw      800h
         end     start
```

In constructing this program it has been assumed that the version for running will be named VIDEO.EXE and that it will be on the disk in the default drive.

The length of the file can be comfortably encompassed in a word so that only a word needs to be reserved for it. Also, the return in the high-order word DX can be ignored when calculating the length by means of function 42H. The length has to be converted to multiples of 16 bytes for system call 31h. This is accomplished by shifting right 4 places. In case the bits lost were non-zero a safety margin is added in afterwards.

The error message is of the simplest kind and not as informative as it should be. It is positioned next to the data segment so that, in a more sophisticated version, it would be relatively easy to see which message was issued for each error. Since it is reached by a JMP from OVER there would be no problem in adding another subroutine, perhaps to retrieve the copy of the screen from the memory, immediately after the RETF of SUBRT.

The whole of the program, including the setting up, remains permanently in the memory. If space were at a premium it would be necessary to have more complex instructions so that only the subroutine was kept.

The subroutine itself is basically a simple copying of the video buffer one word at a time. The complications that occur are due to the segment registers DS, ES, SS all being set to BASIC's data segment when a call is made (CS should be properly adjusted by the BASIC program). Since we want DS to reference the subroutine's data segment and ES to point to the video buffer, arrangements have to be made to preserve and restore DS,ES. Standard linking, as described in Section 4.5, is employed for this process.

Notice that a CALL in BASIC is always a far call i.e. to another segment so that to go back RETF is the right instruction. It would probably be an improvement to replace it by one of the type RETn because one of this kind is relevant when parameters are being passed between BASIC and the assembly program.

Giving the data segment a name makes it easy to point DS to the correct place for the purposes of the subroutine. By attaching PARA we ensure that the four lowest bits of the address of SUBDAT are zero. This means that the address can be treated as a segment with no offset. Furthermore, since SUBDAT is the name of a data segment, the instruction MOV copies the segment part of the address of SUBDAT to AX, whence it is moved to DS. In this way we ensure that DS contains the correct segment value for reference to our data.

The BASIC code to invoke a copy of the screen could be

```
850  PRINT CHR$(27)+"H"
860  DEF SEG=0
870  VIDOFF=PEEK(&H200)+(256*PEEK(&H201))
880  VIDSEG=PEEK(&H202)+(256*PEEK(&H203))
890  DEF SEG=VIDSEG
```

```
900  CALL VIDOFF
910  DEF SEG
```

Instruction 850 is precautionary. It puts the cursor in the home position. There is then no possibility that scrolling could occur while you are trying to copy the screen otherwise the record of the display could be bizarre.

860 tells BASIC to use the segment 0000 for machine language operations like PEEK, POKE, and CALL. The next PEEK then looks at the byte at the offset (in the argument of PEEK) from this segment. Thus 870 recovers the word placed at 0000:0200 by our earlier program; it is consequently the offset of the address of SUBRT. Similarly, 880 supplies the segment of SUBRT.

Instruction 890 directs BASIC to use the segment of SUBRT and then 900 takes the program to the offset of SUBRT within that segment.

When the assembly program has completed its task the return is to 910 which informs BASIC that the default value for a segment should be used from now on.

No parameters are passed in this program. The BASIC CALL does permit the inclusion of parameters. If they are present, the offsets of their addresses are placed on the stack in the order of the parameters. They can be invoked by the machine language program by means of BP as in the linking programs of Chapter 4. A word of warning is justified. It is preferable to make the parameters integers if you can. Anything else entails more intricate programming. In particular, if the parameter is a string the address passed does not point to the string. Instead it transmits the address of a 3 byte *string descriptor* which specifies the length of the string and its address in the string space of BASIC. The content of a string, but not its length, can be altered by a subroutine. Under no circumstances should the string descriptor be changed by an assembly subroutine. If you do aim to change a string in a subroutine you should observe that it may affect the BASIC program unless you deploy the format

```
50  A$="STRING"+""
60  CALL SUBRT(A$)
```

whereupon BASIC copies the string into string space where it can be safely modified without affecting the original text.

9.10 Assembling

The rudiments of the assembly process have already been mentioned in Section 1.11 and now a little more flesh will be put on the bones by referring to the program in the preceding section.

First a source file is fabricated, by means of a word processor or editor, from the instructions and comments as set out in Section 9.9. This file has to have a name which we shall take to be VIDEO. It is normal to attach a filetype which indicates that it is the source for a program in assembly language so it will be VIDEO.A86 or VIDEO.ASM, depending on the system that you are working with. When complete, the file of source code is stored on disk.

Next, the assembler is invoked. Its primary purpose is to convert the source code to machine language. If the source is not on the default drive you will have to add the drive specification to the filename in the usual way (or supply a search path via PATH). In addition to the file of object code the assembler is likely to produce a number of other files, some of which may be optional. Generally, these files will have the same name as, but different extensions from, the source file though some assemblers permit you to change the name. (My personal preference is to keep the name unaltered for ease of identification.) The two most important that you will encounter are VIDEO.OBJ, which contains the object code, and VIDEO.LST which gives a listing. A sample listing for the beginning of our program is:

```
                              SOURCE: VIDEO.A86
                                         PAGE 1

                       cseg
0000  1E        start: push  ds               ; save DS temporarily
0001  BA5700 R         mov   dx,offset subrt  ; get offset of subroutine
0004  B80000 R         mov   ax,seg subrt     ; move segment of
                                                subroutine
0007  8ED8             mov   ds,ax            ; to DS
0009  B080             mov   al,80h           ; insert interrupt number
000B  B425             mov   ah,25h           ; transfer subroutine
                                              ; address
000D  CD21             int   21h              ; to vector table
```

The first column identifies each instruction by giving, in hexadecimal, the number of bytes that have been absorbed in the machine language by the instructions that have gone before. To put it another way, it specifies the value of the location counter at the commencement of an instruction. The second column lists in hexadecimal the object code that results from the instruction on the same line. The third column carries a marker when there is a reference to some other part of the program. The remaining columns are copies of the source file.

The assembler, apart from manufacturing or suppressing individual files, will probably allow you to choose to send the output to a specific disk, to the console or to the printer. Some people prefer to inspect the listing before committing any output to disk. A word of caution is pertinent in this context. The disk must have sufficient space to accommodate not only the files from the assembler but also those after linking. (It is certainly not a good idea to delete any before your routine is thoroughly tested.) At first sight this does not seem much of a problem with a program consisting of a few short instructions.

However, if there is an instruction in the data segment reserving 25 k bytes of memory for some purpose the machine code has to make this provision and therefore the executable file is going to occupy at least 25 k of disk space. (To avoid this there are system calls, not described in this book, for allocating and releasing memory after loading.)

While processing the source program the assembler displays any syntax errors that it has detected. The message usually signals where the error is and identifies it by a numerical code or an English phrase or both. These error messages must be treated as crude guidelines because, for instance, the assembler cannot distinguish between a name which you have forgotten to define and one that you have defined previously but misspelt. Since correction of an error may alter radically the machine code it is wisest to go back to the source program and re-edit before re-assembling, though sometimes you can effect the desired change via a debugger if you are very careful.

Once all syntax errors have been removed it is time to apply the linker. Its job is to make an executable or run file from your object file. Should you wish to join your program to other object modules or library routines which have already been assembled or compiled this is also the responsibility of the linker. Again, you will have the option of selecting a name for the run file. In our case we propose no change and the output of the linker will be VIDEO.EXE or VIDEO.CMD depending on the system (we have assumed the former in Section 9.9). When you are ready to run the program, ignore the extension and just type VIDEO. With luck it will operate as planned but, if not, you will face the tiresome task of finding out what has gone wrong.

In tracing for faults you may solicit the services of a specialist debugger i.e. a utility exclusively designed for examining the content and working of an assembly program. Discussion of such devices, which can differ from each other substantially, is best left to their attendant manuals. There is, however, one aspect which can be confusing with all of them until you realize what is going on. You can expect to find on any debugger an instruction which *disassembles* or *unassembles* i.e. translates machine code back into source code (without any comments or assembler directives, of course, but with the effect of any directives taken into account). There will also be an instruction to display the content of memory with the ASCII equivalent (where it exists) at the side (expect different displays when you examine memory by bytes and by words). Now the machine code may have the same symbol for an instruction and a piece of data; the interpretations are kept separate by having one in the code segment and the other in the data segment. Should you ask, either fortuitously or deliberately, the debugger to disassemble some data it will attempt the translation into source code and generate some instructions which are unrecognisable as part of your program. Likewise, if you display the content of code segment addresses you are likely to think that you have some inexplicable data in your program. Therefore, before disassembling some code or displaying some data you should verify where the code and data segments are located (there should be an instruction which conveys this information). There is a related point—if your disassembler allows you to start with any address you must make sure that you choose one which is the beginning of an instruction otherwise the translation may result in some unexpected (and fictitious) source code.

9.11 Summary of system calls

A short list of the system calls is laid out below

Number	Function	Section
01H	Read keyboard and echo	9.2
02H	Display character	9.2
05H	Print character	9.2
08H	Read keyboard	9.2
09H	Display string	9.2
0AH	Buffered input	9.2
0FH	Open file	9.4
10H	Close file	9.4
14H	Sequential read from disk	9.5
15H	Sequential write to disk	9.5
16H	Create file	9.4
1AH	Set disk transfer address	9.5
25H	Set vector address	9.9
31H	Terminate but leave resident	9.9
35H	Get interrupt vector	9.3
3CH	Create handle	9.8
3DH	Open handle	9.8
3EH	Close handle	9.8
3FH	Read file or device	9.7
40H	Write file or device	9.7
42H	Move read/write pointer	9.9
4CH	End process	9.6

9.12 Overlays

Large programs can make heavy demands on the space in memory; in fact, if they require more than 64 k of code or data they cannot be mounted at all. However, it may not be necessary that a large program resides in memory all at once, especially if you have employed a modular design. For instance, many programs are controlled by *menus* i.e. at certain stages the user is offered a number of options to choose from. Each option allows the user to perform a different function. Since the functions are distinct and the user never invokes more than one of them at a time, they do not all have to reside in memory simultaneously. Moreover, when the task has been completed, the program returns to the menu ready for the next decision by the user. Consequently, each option in the menu can be allocated its own subprogram which can be stored on disk and loaded only when it is the chosen one. Furthermore, since the subprograms are never in the memory simultaneously there is no reason why they should not be loaded into the same memory location.

Subprograms which share memory locations are known as *overlays*. Overlays provide a mechanism for economizing on the space in memory. While their value for confining large programs to a limited space has already been explained they may also be helpful with programs of moderate size when the working space proves to be inadequate in applications.

An overlay is constructed as a module in the normal way and ends with the instruction RET. Therefore, one method of incorporating an overlay is by a standard CALL. For a simple example, we write a program which displays 'In base now' and then, by means of an overlay, displays 'Overlay done'. (Note: the detailed technique for embracing overlays varies with the assembler and the linker)

```
; base.a86   : to display certain messages

             cseg
             extrn   overlay1:near         ; OVERLAY1 is defined
                                           ; in another program
base:        mov     dx,offset basem       ; address of string to DX
             mov     ah,09h                ; code to display string
             int     21h                   ; display string
             mov     dx,offset ovlaym      ; address of string to DX
             call    overlay1              ; call the overlay
             ret                           ; back to calling program
;
             dseg
basem        db      'In base now',0dh,0ah,'$'
ovlaym       db      'Overlay done',0dh,0ah,'$'
             end
```

For the overlay we have

```
; overlay1.a86: an overlay to display a message
;
             cseg
overlay1:    mov     ah,09h                ; code to diaplay string
             int     21h                   ; display string
             ret                           ; return to BASE
```

The advantage of this method is that, because it is achieved by a standard CALL, a firm decision on whether you want an overlay or not can be deferred until linking. If you decide on an overlay then OVERLAY1.OVR will be created to carry out the duties set out for the overlay.

When BASE runs it first displays 'In base now' on the console. As soon as it reaches the CALL instruction it transfers control to the Overlay Manager. The Overlay Manager loads OVERLAY1.OVR from the default drive and hands over control to it, unless the requested overlay is already in memory when it is not reloaded before control is transferred.

The overlay then displays 'Overlay done' on the monitor. As soon as it is finished, control passes to the instruction immediately after the CALL instruction in BASE. The program continues from that point in the usual manner.

There are a couple of disadvantages with this method. Firstly, since the label in the CALL instruction must agree with the name of the OVR file, the names of overlays are fixed. If you want to change the name of an overlay you must edit, reassemble, and relink. Secondly, the Overlay Manager loads the overlay only from the drive which was the default drive when the program which calls the overlay began execution. Any changes in drive which occur after the commencement of execution are disregarded.

For these reasons a more complicated procedure has been devised in which the Overlay Manager is addressed directly. The technique will be illustrated by going through the previous example again. Since the instructions of OVERLAY1.A86 are not affected they will not be repeated.

```
; base1.a86:        to display certain messages
;
                    cseg
                    extrn   ?ovlay:near         ; declaration of entry
                                                ; to Overlay Manager
  base1:            mov     dx,offset basem     ; address of string to DX
                    mov     ah,09h              ; code to display string
                    int     21h                 ; display string
;
                    mov     dx,offset ovlaym    ; address of string to DX
                    call    ?ovlay              ; call Overlay Manager
                    dw      overlay_name        ; offset of name of overlay
                    db      0                   ; clear load flag
                    ret                         ; finished
;
                    dseg
  basem             db      'In base now',0dh,0ah,'$'
  ovlaym            db      'Overlay done',0dh,0ah,'$'
  overlay_name      db      'OVERLAY1   '       ; name of
                                                ; overlay to be loaded
                    end
```

The principal differences are in EXTRN and the CALL instructions. The EXTRN identifies the Overlay Manager so that the CALL can refer to it.

The two directives which follow immediately after the CALL are obligatory. The first specifies the offset of the name of the overlay to be loaded. The actual name is declared in the data segment in a string of ten characters. In this case the name of the overlay (without extension) has been padded with two blanks at the end to make it up to ten characters. The Overlay Manager will look for the overlay on the default drive according to the same rule as enunciated in the first method. But the Overlay Manager can be forced to try a particular drive, say drive B, by altering the directive in the data segment to

```
overlay_name    db    'B:OVERLAY1'
```

Thus the two extra characters in the name string are to permit you the possibility of prescribing the drive. Hence the second method for overlays offers more freedom than the first.

The second directive after CALL is to adjust the *Load Flag*. When it is cleared, as in the example, the Overlay Manager knows that the overlay is to be loaded only if it is not in memory already. The directive

```
db    1
```

sets the Load Flag to 1 and then the Overlay Manager loads the designated overlay irrespective of whether it is already in memory.

The program BASE1 behaves in exactly the same way as BASE up to the CALL instruction. The CALL (with its two associated directives) advises the Overlay Manager to load OVERLAY1.OVR from the default drive and transfer control to it. When OVERLAY1 is finished control returns to BASE1 in the customary fashion.

Although OVERLAY_NAME has been assigned as OVERLAY1 by the data segment you could arrange to furnish it as a character string from some other source such as the keyboard.

Nesting of overlays is generally feasible though there may be a constraint, such as 5, on the number of levels of nesting. You may also find that the Overlay Manager uses the default buffer at 80H in the Data Segment so that it is advisable to keep clear of this area for your own data.

9.13 Structured programming

No really large or complex programs have been constructed in this book. Nevertheless, the principles which underlie the construction of such programs are no different from those which have been illustrated throughout in the preparation of our smaller programs. Be as simple and clear as you can. Use plenty of comments so that if you (or someone else) come back to it later the task of understanding it is as straightforward as possible. In particular, ensure that the rules for any interfaces are plainly stated.

With a complicated task carry out the broad strategy first. Concentrate on the duties of each main part and how the parts are to be linked before spending too much time on the detailed operation of each part. If your interfaces are soundly based on a good overall strategy you can have a working program available for use while, at your leisure, you optimize the detail of a particular part should you decide that to be desirable. In splitting up the major job into its components do not forget to take advantage of any routines that you already have that are working well and adapt the structure to incorporate them. Unnessary repetition wastes time that you could be devoting to the snags that are bound to arise.

In other words, complex programs are built by combining programs, preferably ones that exist already, so that the programs are effectively regarded as modules. The interfaces connecting the modules should be kept as elementary as feasible yet be sufficiently robust so that the inner working of a module can be tinkered with without throwing the whole system awry. This could be restated as modules should be designed on a 'need to know' basis i.e. each performs its own function with the minimum awareness of what the rest are doing. Each, of course, will have its own error traps and, ideally, will be able to accommodate a parameter change without having to rewrite numerous lines scattered through the entire module.

Some people refer to the job of resolving a problem for the computer along these lines as *structured programming*. The aim is to achieve reliable, easily-comprehensible, code which is composed of sections which work smoothly together but can be handled internally virtually independently. The life of programmers and users is much happier when this objective can be attained.

As a final word we mention that the implementation of the principles enunciated above can be seen, on a small scale, in the program of Section 9.6. Names are defined in the data segment for things like the buffer size and only the symbolic name appears in the code segment. A single change in the data segment will alter the buffer size and the new size will be inserted automatically in the code segment at the right places. The operation of obtaining a character from the buffer is broken down into three modules, each fulfilling its own function and only called on when required. The error traps are obvious. The common use of a part of the program by other parts is exemplified by calling on BDOS to issue an interrupt. Actually, if we had not wished to make this point, a more economical program would have resulted from employing a direct instruction to interrupt instead of a CALL. One of the judgements a programmer may have to make is to weigh economy against flexibility and robustness.

Exercises

1. Write a program that prints 80 consecutive ASCII characters starting with ! and then follows with a carriage return and line feed. On the next line do the same but start at the next character along, i.e. ". Continue doing this until a line

would contain an unprintable character when the program should terminate.
2. Do the same as in Exercise 1, but repeat the pattern four times before ending the process. Would pressing CONTROL-C during the execution of the program have any effect?
3. Write a program which displays a key when it is pressed and then displays its hexadecimal code.
4. Write a program which simultaneously displays and prints the character from a key which is pressed. If carriage return is entered both the screen and printer are to undertake a carriage return and line feed.
5. An ASCII file MYFILE.TXT is stored on the disk in drive B. Write a program to print the characters in it.
6. Write a program to display in hexadecimal the contents of a file on disk (not necessarily stored in ASCII format).
7. A sequence of 64 consecutive occurrences of the letter A has been placed somewhere in memory. Write a program to locate the address of the start of this sequence and display it on the screen.
8. An ASCII disk file contains a sequence of strings of variable length, each string being terminated by a carriage return and line feed. The strings are to be sorted alphabetically in the sense that characters lower in ASCII are regarded as preceding those which are higher. The output is to go to STDOUT. Construct a program based on bubblesort.
9. To speed up bubblesort in arranging an array of integers into ascending order it is suggested that elements which are far apart should be compared instead of adjacent ones, at any rate at the beginning. For the new method, called *shell sort*, the following is proposed to you as a possible Pascal program

```
procedure shell (var v: intarray; n: integer);
var
        gap, i, j, jg, k: integer;
begin
        gap := n div 2;
        while (gap>0) do begin
        for i :=gap+1 to n do begin
        j := -gap;
        while (j>0) do begin
        jg :=j+gap
        if (v[j]< =v[jg]) then j :=0
        else begin
        k :=v[j];
        v[j] :=v[jg];
        v[jg] :=k
        end;
        j :=j-gap
        end
        end;
```

```
gap := gap div 2
end
end;
```

After making any alterations you consider desirable construct an assembly program which achieves the same objective, deciding whether or not you want to include STDIN and/or STDOUT.

10. Write an assembly program to clear the screen and put the cursor in a specified spot by invoking escape strings.

11. Devise a program IDENTIFY which displays the message 'Please give your name', accepts your name (up to 40 characters) as input and then displays your name preceded by 'Welcome to'.

Add the extra facility that the name inserted is compared with a specified name and, if they agree, the welcome is issued but, when there is disagreement, the message 'Sorry! You are unknown' is displayed followed by a repeat request for the name. (A further level of sophistication would be to allow a user to change the specified name once his name had been accepted.)

12. A string, which contains not more than 30 bytes, is to be accepted from the keyboard. The string is to be copied to the printer as many times as are necessary to completely fill five lines of print. Devise a suitable program.

13. The file RECORD.TXT is on the disk in drive A. The file contains 26 records each of length 20 bytes. The user is asked for a letter of the alphabet. The letter is then converted to a numerical scale by the code A = 1, B = 2, and so on. The record with that number is then displayed. Write a program to achieve this. (Do not forget a carriage return and line feed after displaying the record before issuing the prompt again.)

14. Each member of a course is asked to supply a name (not more than 30 characters) and a home town (not more than 20 characters). After entry this information is to be displayed on the same horizontal line followed by an identifying number starting at 1. If the record is confirmed it is to be placed in the file COURSE.REC on the disk in drive B. The procedure ceases on entering ZZZZ. Write an assembly program for this purpose.

15. Write a program which prompts for the number of a member of the course in Exercise 14 and then prints the corresponding record from COURSE.REC.

16. The copy of the video buffer saved in memory in Section 9.9 is to be placed on disk instead. Construct the assembly program and accompanying BASIC instructions.

17. Write the BASIC routine and assembly subroutine that will retrieve the copy of the video buffer in Section 9.9 from memory and display it on the monitor.

18. While running a BASIC program you wish to be able, at a certain stage, to examine a small HELP window and then return to the display immediately before you called up the window. Devise programs that will achieve this.

REFERENCES

For more advanced methods and technical detail see

iAPX86 User's Manual, Programmers Reference, published by Intel.
IBM Macro Assembler published by IBM.
MS-DOS Programmer's Reference Manual published by Microsoft.
Programmer's Utilities Guide for the CP/M86 published by Digital Research.

APPENDIX A: ASCII CODES

In the following table the high-order bits are shown only when they change. The first 32 entries describe codes for control or communications. The codes from 127 onwards may differ from system to system.

Deci-mal	Hexa-decimal	Binary		Symbol	Deci-mal	Hexa-decimal	Binary		Symbol
		High-order bits	Low-order bits				High-order bits	Low-order bits	
0	00H	0000	0000	Null	32	20H	0010	0000	Space
1	01H		0001	Start of heading	33	21H		0001	!
2	02H		0010	Start of text	34	22H		0010	”
3	03H		0011	End of text	35	23H		0011	#
4	04H		0100	End transmission	36	24H		0100	$
5	05H		0101	Enquiry	37	25H		0101	%
6	06H		0110	Acknowledge	38	26H		0110	&
7	07H		0111	Bell	39	27H		0111	’
8	08H		1000	Backspace	40	28H		1000	(
9	09H		1001	Horizontal tab	41	29H		1001	)
10	0AH		1010	Line feed	42	2AH		1010	*
11	0BH		1011	Vertical tab	43	2BH		1011	+
12	0CH		1100	Form feed	44	2CH		1100	,
13	0DH		1101	Carriage return	45	2DH		1101	–
14	0EH		1110	Shift out	46	2EH		1110	.
15	0FH		1111	Shift in	47	2FH		1111	/
16	10H	0001	0000	Data link escape	48	30H	0011	0000	0
17	11H		0001	Device control 1	49	31H		0001	1
18	12H		0010	Device control 2	50	32H		0010	2
19	13H		0011	Device control 3	51	33H		0011	3
20	14H		0100	Device control 4	52	34H		0100	4
21	15H		0101	Negative acknowledge	53	35H		0101	5
22	16H		0110	Synchronous idle	54	36H		0110	6
23	17H		0111	End transmission block	55	37H		0111	7
24	18H		1000	Cancel	56	38H		1000	8
25	19H		1001	End of medium	57	39H		1001	9
26	1AH		1010	Substitute	58	3AH		1010	:
27	1BH		1011	Escape	59	3BH		1011	;
28	1CH		1100	File separator	60	3CH		1100	<
29	1DH		1101	Group separator	61	3DH		1101	=
30	1EH		1110	Record separator	62	3EH		1110	>
31	1FH		1111	Unit separator	63	3FH		1111	?

Decimal	Hexadecimal	Binary		Symbol	Decimal	Hexadecimal	Binary		Symbol
		High-order bits	Low-order bits				High-order bits	Low-order bits	
64	40H	0100	0000	@	96	60H	0110	0000	
65	41H		0001	A	97	61H		0001	a
66	42H		0010	B	98	62H		0010	b
67	43H		0011	C	99	63H		0011	c
68	44H		0100	D	100	64H		0100	d
69	45H		0101	E	101	65H		0101	e
70	46H		0110	F	102	66H		0110	f
71	47H		0111	G	103	67H		0111	g
72	48H		1000	H	104	68H		1000	h
73	49H		1001	I	105	69H		1001	i
74	4AH		1010	J	106	6AH		1010	j
75	4BH		1011	K	107	6BH		1011	k
76	4CH		1100	L	108	6CH		1100	l
77	4DH		1101	M	109	6DH		1101	m
78	4EH		1110	N	110	6EH		1110	n
79	4FH		1111	O	111	6FH		1111	o
80	50H	0101	0000	P	112	70H	0111	0000	p
81	51H		0001	Q	113	71H		0001	q
82	52H		0010	R	114	72H		0010	r
83	53H		0011	S	115	73H		0011	s
84	54H		0100	T	116	74H		0100	t
85	55H		0101	U	117	75H		0101	u
86	56H		0110	V	118	76H		0110	v
87	57H		0111	W	119	77H		0111	w
88	58H		1000	X	120	78H		1000	x
89	59H		1001	Y	121	79H		1001	y
90	5AH		1010	Z	122	7AH		1010	z
91	5BH		1011	[	123	7BH		1011	{
92	5CH		1100	\	124	7CH		1100	¦
93	5DH		1101	]	125	7DH		1101	}
94	5EH		1110	^	126	7EH		1110	~
95	5FH		1111	_	127	7FH		1111	⌂

The binary representation of the following is omitted but it should be noted that they all have 1 in the most significant position.

Decimal	Hexa-decimal	Symbol	Decimal	Hexa-decimal	Symbol
128	80H	Ç	160	A0H	á
129	81H	ü	161	A1H	í
130	82H	é	162	A2H	ó
131	83H	â	163	A3H	ú
132	84H	ä	164	A4H	ñ
133	85H	à	165	A5H	Ñ
134	86H	å	166	A6H	a
135	87H	ç	167	A7H	o
136	88H	ê	168	A8H	¿
137	89H	ë	169	A9H	⌐
138	8AH	è	170	AAH	¬
139	8BH	ï	171	ABH	½
140	8CH	î	172	ACH	¼
141	8DH	ì	173	ADH	¡
142	8EH	Ä	174	AEH	≪
143	8FH	Å	175	AFH	≫
144	90H	É	176	B0H	░
145	91H	æ	177	B1H	▒
146	92H	Æ	178	B2H	▓
147	93H	ô	179	B3H	│
148	94H	ö	180	B4H	┤
149	95H	ò	181	B5H	╡
150	96H	û	182	B6H	╢
151	97H	ù	183	B7H	╖
152	98H	ÿ	184	B8H	╕
153	99H	Ö	185	B9H	╣
154	9AH	Ü	186	BAH	║
155	9BH	¢	187	BBH	╗
156	9CH	£	188	BCH	╝
157	9DH	¥	189	BDH	╜
158	9EH	Pt	190	BEH	╛
159	9FH	ƒ	191	BFH	┐

Decimal	Hexa-decimal	Symbol	Decimal	Hexa-decimal	Symbol
192	C0H	└	224	E0H	α
193	C1H	┴	225	E1H	β
194	C2H	┬	226	E2H	Γ
195	C3H	├	227	E3H	π
196	C4H	─	228	E4H	Σ
197	C5H	┼	229	E5H	σ
198	C6H	╞	230	E6H	μ
199	C7H	╟	231	E7H	τ
200	C8H	╚	232	E8H	Φ
201	C9H	╔	233	E9H	Θ
202	CAH	╩	234	EAH	Ω
203	CBH	╦	235	EBH	δ
204	CCH	╠	236	ECH	∞
205	CDH	═	237	EDH	φ
206	CEH	╬	238	EEH	∈
207	CFH	╧	239	EFH	∩
208	D0H	╨	240	F0H	≡
209	D1H	╤	241	F1H	±
210	D2H	╥	242	F2H	≥
211	D3H	╙	243	F3H	≤
212	D4H	╘	244	F4H	⌠
213	D5H	╒	245	F5H	⌡
214	D6H	╓	246	F6H	÷
215	D7H	╫	247	F7H	≈
216	D8H	╪	248	F8H	°
217	D9H	┘	249	F9H	•
218	DAH	┌	250	FAH	-
219	DBH	█	251	FBH	√
220	DCH	▄	252	FCH	^{n}
221	DDH	▌	253	FDH	^{2}
222	DEH	▐	254	FEH	■
223	DFH	▀	255	FFH	(blank 'FF')

APPENDIX B: 8086 MNEMONICS*

The following list shows the mnemonics in capitals for clarity. In preparing a program either lower or upper case characters can be employed since the assembler will automatically convert them to upper case if necessary.

Each mnemonic is followed by a brief description of its use, any restrictions on its applicability, its effect on flags and a rough measure for its time of operation in cycles. Since times can vary considerably according as operands are memory or register the timings should be regarded as crude guides at best. The status flags are shown in the order OSZAPC, the letters corresponding to those of Section 2.3. A letter indicates that the flag is raised or lowered as appropriate while a dash means that the flag is not affected. If an operation forces a flag to be 1 or 0 whatever the circumstances this is shown explicitly. A ? signifies that the result of an operation is undefined.

As regards timings, precise values can vary widely according as registers, memory, indirect addresses, etc. are involved. To avoid a profusion of detail only rough orders of magnitude for operations are given with r, m, b, w indicating registers, memory, bytes, words for noteworthy differences. For exact information covering all possibilities consult Intel.

In the following the term register excludes a segment register unless the contrary is stated. If s or d is unspecified it can be either register or memory. The section number in the right hand column shows where the instruction is first encountered in the main text.

Mnem-onics	Syntax	Description	Cycles	Flags OSZAPC	Sec-tion
AAA	aaa	Adjust AL after unpacked BCD addition. If right four bits exceed 9 or AF = 1 add 6 to them; then put AH = 1, CF = 1 and clear left four bits of AL.	4	???A?C	6.6
AAD	aad	Adjust AX *before* unpacked BCD division. Multiply AH by 10 and add it to AL. Clear AH.	60	?SZ?P?	6.8

* All mnemonics are the copyright of the Intel Corporation 1986

Mnem-onics	Syntax	Description	Cycles	Flags OSZAPC	Sec-tion
AAM	aam	Adjust AL after unpacked BCD multiplication. Divides AL by 10, putting quotient in AH and remainder in AL.	83	?SZ?P?	6.7
AAS	aas	As AAA but for subtraction.	4	???A?C	6.6
ADC	adc d,s	Same conditions as ADD but CF is added to result	r3,m25	OSZAPC	4.2
ADD	add d,s	Add the content of s to the content of d leaving result in d. One at least of s and d must be a register or s must be an immediate value	r3,m25	OSZAPC	2.5
AND	and d,s	Combine bits by logical AND rules of Section 1.5(b). If neither s nor d is a register, s must be an immediate value.	r3,m25	0SZ?P0	6.1
CALL	call d	Push next instruction address on stack and jump to d, if it is a label, or content of d if it is memory or 16 bit register.	19	------	4.2
CALLF	callf d	As CALL but CS is placed on stack first. For call to another segment (not available on all assemblers).	30	------	
CBW	cbw	Convert byte in AL to word in AX with sign extension	2	------	2.10
CLC	clc	CF=0.	2	-----0	6.2
CLD	cld	Clear DF, causing string instructions to auto-increment the operand registers.	2	------	5.2
CLI	cli	Make IF=0, so that interrupts are disabled or masked.	2	------	6.2
CMC	cmc	Perform ones-complement of CF.	2	-----C	6.2
CMP	cmp d,s	Form d-s but adjust only flags. If s is not an immediate value one of s and d must be a register.	r3, m18	OSZAPC	2.14

Mnem-onics	Syntax	Description	Cycles	Flags OSZAPC	Sec-tion
CMPS	cmps dummy, dummy	Subtract 1 byte (or word) of destination [DI] from source [SI], adjust flags but do not return result. Update SI and DI.	22	OSZAPC	5.9
CMPSB	cmpsb	The same as CMPS but for byte operand only.		OSZAPC	5.9
CMPSW	cmpsw	The same as CMPS but for word operand only.		OSZAPC	5.9
CWD	cwd	Convert word in AX to doubleword in DX, AX with sign extension.	5	------	2.10
DAA	daa	Adjust AL, after addition, to packed BCD format. If right four bits exceed 9 or AF=1 add 6 to them. If left four bits exceed 9 or CF=1 add 6 to them.	4	?SZAPC	6.5
DAS	das	As DAA but for subtraction.	4	?SZAPC	6.5
DEC	dec d	Decrease the content of d by 1. d cannot be immediate value.	r2, m20	OSZAP–	3.1
DIV	div s	Divide (unsigned) DX! AX (word) or AX (Byte) by content of s. If byte results, AL= quotient, AH=remainder. If word results, AX=quotient, DX=remainder. s cannot be immediate value.	b100, w170	??????	2.10
ESC		Used by 8087 coprocessor.	r2, m17	------	
HLT	hlt	Enter halt state until an interrupt is recognized.	2	------	7.4
IDIV	idiv s	Same as DIV but for signed numbers.	b130, w200	??????	2.8
IMUL	imul s	If s is (signed) byte multiply by AL and place result in AH, AL. If s is word multiply by AX and place result in DX,AX. s cannot be immediate value.	b110, w170	O????C	2.7

Mnemonics	Syntax	Description	Cycles	Flags OSZAPC	Section
IN	in d,s	Transfer byte (d = al) or word (d = ax) from port. The number of the port is s, if s is an immediate value (0–255), or contained in DX if s = dx.	b10, w8	------	7.5
INC	inc d	Increase the content of d by 1. d cannot be immediate value.	r2, m20	OSZAP–	2.14
INT	int s	s is interrupt vector number (0–255). Push flags on stack, clear TF and IF flags, push CS then IP to stack. First word of interrupt pointer to IP, second word to CS.	50	------	9.2
INTO	into	If OF = 1 same as INT for interrupt vector 4 (location 10H) but ignored if OF = 0	50 (OF = 1) 4 (OF = 0)	------	—
IRET	iret	Restore IP, CS, flags (in that order) from stack and return control to address saved by interrupt.	24	Set from stack	7.4
JCXZ	jcxz d	Jump, if CX register is zero, to label d within ± 128 bytes	12	------	3.5
J.	j. d	A conditional jump to a label d within ± 128 bytes. The . stands for letters set out in Appendix C.	10	------	2.13
JMP	jmp d	Jump to d if a label. If d is memory or 16 bit register jump to location within.	30	------	2.13
JMPF	jmpf d	Far jump to d which must be label possibly in another code segment. (Subsumed in JMP in some assemblers)	30	------	2.13
JMPS	jmps d	Short jump to label within ± 128 bytes. (Subsumed in JMP in some assemblers)	30	------	2.13
LAHF	lahf	Transfer flags to AH, bits in AH being SZ?A?P?C.	4	------	6.2

Mnem-onics	Syntax	Description	Cycles	Flags OSZAPC	Sec-tion
LDS	lds d,s	Transfer first word of s (which must contain double word) to d (which must be 16 bit register) and second word to DS.	25	------	7.3
LEA	lea d,s	Transfer the (offset) address of s to d, which must be a 16 bit register.	10	------	3.2
LES	les d,s	Transfer first word of s (contains double word) to d (which must be 16 bit register) and second word to ES.	25	------	7.3
LOCK	lock	Locks the bus. It can be a prefix to any other instruction and prevents access to control of the bus by others for the duration of the operation resulting from the instruction.	2	------	—
LODS	lods dummy	Move 1 byte (or word) from source [SI] to AL (or AX) and update SI.	12	------	—
LODSB	lodsb	The same as LODS but only a byte is moved.			
LODSW	lodsw	The same as LODS but only a word is moved.			
LOOP	loop d	Decrease CX by 1. If resulting CX is not zero jump to label d (within ± 128 bytes).	17	------	4.4
LOOPE	loope d	Decrease CX by 1. If resulting CX is not zero and ZF=1 jump to label d (within ± 128 bytes).	18	------	4.4
LOOPNE	loopne d	Decrease CX by 1. If resulting CX is not zero and ZF=0 jump to label d (within ± 128 bytes).	19	------	4.4
LOOPNZ		The same as LOOPNE.		------	4.4
LOOPZ		The same as LOOPE.			

Mnem-onics	Syntax	Description	Cycles	Flags OSZAPC	Sec-tion
MOV	mov d,s	Copy the content of s to d. One at least of s and d must be a register (including a segment register) or s must be an immediate value.	r2, m20	——	2.5
MOVS	movs dummy, dummy	Move 1 byte (or word) from source [SI] to destination [DI], then update SI,DI.	18	------	5.4
MOVSB	movsb	The same as MOVS but only a byte is moved.			5.4
MOVSW	movsw	The same as MOVS but only a word is moved.			5.4
MUL	mul s	Same as IMUL but for unsigned magnitudes.	b90, w150	O????C	2.10
NEG	neg d	Changes the sign of d by performing twos-complement. If d=0 puts CF=0.	r3, m20	OSZAP1	3.3
NOP	nop	No operation is performed.	3		1.5
NOT	not d	Changes every 1 to 0 and every 0 to 1.	r3, m20	------	6.1
OR	or d,s	Combine bits by logical OR rules of Section 1.5(b). If neither s nor d is a register, s must be an immediate value.	3	0SZ?P0	6.1
OUT	out d,s	Transfer byte (s=al) or word (s=ax) to port. The number of the port is d, if d is an immediate value (0–255), or contained in DX if d=dx.	b10, w8	------	7.6
POP	pop d	Move the top stack element (word) to d (which can be segment register).	r8, m25	------	4.3
POPF	pop f	Transfer top stack element to flags	8	Set by stack	6.2
PUSH	push s	Move the word in s (which can be segment register) onto stack.	r11, m25	------	4.3
PUSHF	pushf	Copy flags to top stack element.	10	------	6.2

Mnem- onics	Syntax	Description	Cycles	Flags OSZAPC	Sec- tion
RCL	rcl s,1	Rotate 1 bit left through CF. OF = 1 if sign bit changed.	r2, m24	O----C	—
	rcl s,cl	Rotate left through CF the number of bits in CL.	r8 + 4CL, m30 + 4Cl	?----C	—
RCR		The same as RCL but rotation to the right.			—
REP		Prefix to a string instruction. Decreases CX by 1, and, if then CX ≠ 0, repeats the string instruction.	2	------	5.4
REPE		The same as REP but with repetition only if CX ≠ 0 and ZF = 1.			5.4
REPNE		The same as REP but with repetition only if CX ≠ 0 and ZF = 0.			5.4
REPNZ		The same as REPNE.			5.4
REPZ		The same as REPE.			5.4
RET	ret	Return to address pushed by previous CALL. Increase SP by 2.	8	------	4.2
	ret number	As RET but SP increased by 2 + number.	12	------	4.5
RETF	retf	As RET but SP increased by 4 or 4 + number. For return from another segment.	18	------	9.9
	retf number		17	------	9.9
ROL	rol d,1	Same as ROR but rotation to left.	r2, m25	O----C	6.1
	rol d,cl		r8 + 4CL, m30 + 8CL	?----C	6.1
ROR	ror d,1	Rotate in a circle to right one position. Last bit out left in CF. OF = 1 if sign bit changed.	r2, m25	O----C	6.1
	ror d,cl	Rotate number of positions in CL. Last bit out left in CF.	r8 + 4CL, m30 + 4CL	?----C	6.1
SAHF	sahf	Transfer AH, as SZ?A?P?C, to flags.	4	-SZAPC	6.2

Mnem-onics	Syntax	Description	Cycles	Flags OSZAPC	Sec-tion
SAL	sal d,1	Shift 1 bit left, losing bit shoved out, filling on right with 0. Last bit out left in CF. OF = 1 if sign bit changed.	r2, m25	OSZ?PC	1.5
	sal d,cl	Shift left number of positions in CL. Last bit out left in CF.	r8 + 4CL, m30 + 4Cl	?SZ?PC	1.5
SAR	sar d,1 sar d,cl	Shift right 1 or number in CL, losing bits shoved out, filling on left with copy of sign bit. Last bit out left in CF.	As SAL	OSZ?PC	1.5
SBB	sbb d,s	Same conditions as SUB but CF is subtracted from the result.	r4, m25	OSZAPC	—
SCAS	scas dummy	Subtract [DI] from AL or AX, affect flags but do not return result. Update DI.	15	OSZAPC	5.3
SCASB	scasb	As SCAS but subtract from AL only			5.3
SCASW	scasw	As SCAS but subtracts from AX only.			5.3
SHL	shl d,1 shl d,cl	The same as SAL.			6.4
SHR	shr d,1	Shift 1 bit right, losing bit shoved out, filling on left with 0. Last bit left in CF. OF = 1 if signed bit changed.	As SAL	OSZ?PC	6.11
	shr d,cl	Shift number of positions in CL.		?SZ?PC	6.11
STC	stc	Set CF = 1	2	-----1	6.2
STD	std	Set DF = 1, causing string instructions to autodecrement the operand registers.	2	------	5.6
STI	sti	Set IF = 1, so that interrupts are enabled.	2	------	6.2
STOS	stos dummy	Transfer AL or AX to [DI], then update DI.	11	------	5.1
STOSB	stosb	As STOS but transfers AL only.			5.2
STOSW	stosw	As STOS but transfers AX only.			5.2

Mnem-onics	Syntax	Description	Cycles	Flags OSZAPC	Sec-tion
SUB	sub d,s	Subtract the content of s from contend of d, leaving result in d. If neither s nor d is a register, s must be an immediate value.	r3, m25	OSZAPC	2.6
TEST	test d,s	Logical AND of contents of d and s but only flags affected. If neither s nor d is a register, s must be an immediate value.	r4, m20	0SZ?P0	6.10
WAIT	wait	Processor enters halt state until signal on its TEST pin is asserted.	3	------	—
XCHG	xchg d,s	Exchange the contents of s and d. One at least of s and d must be a register; neither can be an immediate value.	r4, m25	------	6.1
XLAT	xlat table	Copies byte, with offset AL from beginning of table (maximum 256 bytes) pointed to by BX, to AL.	11	------	—
XOR	xor d,s	Combine bits by logical XOR rules of Section 1.5(b). If neither s nor d is a register, s must be an immediate value.	r3, m25	0SZ?P0	2.11

APPENDIX C: CONDITIONAL JUMPS

A conditional jump usually occurs immediately after a comparison,

cmp   d,s

in which d–s is performed and the result is recorded only by its effect on the flags; d and s are left unaltered. In the following,* above and below refer to unsigned magnitudes whereas greater than and less than refer to signed numbers.

JA	Jump if above (CF = 0 and ZF = 0; d > s unsigned)
JAE	Jump if above or equal (CF = 0; d ⩾ s unsigned)
JB	Jump if below (CF = 1; d < s unsigned)
JBE	Jump if below or equal (CF and ZF not both 0; d ⩽ s unsigned)
JC	Jump if CF = 1 (the same as JB)
JE	Jump if equal (ZF = 1; d = s unsigned)
JG	Jump if greater (ZF = 0 and SF = OF; d > s signed)
JGE	Jump if greater or equal (SF = OF; d ⩾ s signed)
JL	Jump if less (SF ≠ OF; d < s signed)
JLE	Jump if less or equal (SF ≠ OF or SF = OF and ZF = 1; d ⩽ s signed)
JNA	Jump if not above (same as JBE)
JNAE	Jump not above nor equal (same as JB)
JNB	Jump if not below (same as JAE)
JNBE	Jump if not below nor equal (same as JA)
JNC	Jump if CF = 0 (same as JNB and JAE)
JNE	Jump if not equal (ZF = 0; d ≠ s)
JNG	Jump if not greater (same as JLE)
JNGE	Jump if not greater nor equal (same as JL)
JNL	Jump if not less (same as JGE)
JNLE	Jump if not less nor equal (same as JG)
JNO	Jump if not overflow (OF = 0)
JNP	Jump if not even parity (PF = 0)
JNS	Jump if not sign (SF = 0)
JNZ	Jump if not zero (same as JNE)
JO	Jump if overflow (OF = 1)
JP	Jump if even parity (PF = 1)
JPE	Jump if parity even (same as JP)
JPO	Jump if parity odd (same as JNP)
JS	Jump if sign set (SF = 1)
JZ	Jump if zero (same as JE)

*All mnemonics copyright Intel Corporation 1986

APPENDIX D: ASSEMBLER DIRECTIVES

This appendix lists some alternative forms for assembler directives that may be encountered.

Text	Description	Alternative	Notes
rb 1	Reserve 1 byte with unspecified value	db 1 ?	
rb 10	Reserve 10 bytes with unspecified values	db 10 dup(?)	
db 3,3,3,3	Declare 4 bytes with the value 3	db 4 dup(3)	May be restricted to code segment
dw 3,3,3,3	Declare 4 words with the value 3	dw 4 dup(3)	May be restricted to code segment
dd 3,3,3,3	Declare 4 double words with the value 3	dd 4 dup(3)	May be restricted to code segment
rw 1	Reserve 1 word with unspecified value	dw 1 ?	Duplicate form available as for db
rd 1	Reserve 1 double word with unspecified value	dd 1 ?	
'Abc'	String	"Abc"	
'$'	End of string for display	"$"	
extrn init: near	INIT declared elsewhere	init extrn	

In the text CSEG, DSEG, ESEG, SSEG indicate segments of code, the first letter showing which segment register is involved. The alternative version avails itself of the generic term SEGMENT, distinguishing different segments by name. Before a new segment can be started the old one must be terminated by an end segment directive ENDS with the name of the segment attached. Thus, you might have

```
chaff    segment
         . . .
chaff    ends
finch    segment
         . . .
finch    ends
```

The assembler is now left unaware of what segment register is associated with each segment of code. Therefore, another directive ASSUME tells the assembler which segment register is relevant. For example, with the above segments you might have

```
assume    cs:finch,ds:chaff
```

Since the assembler has to know what address will be in a segment register when the program is run before it can construct the machine code, the ASSUME directive must precede any instruction that will result in object code. Normally, it is the first instruction in the code segment.

There is a peculiarity of the ASSUME directive which has to be watched. Once it has identified the code segment, the register CS is looked after automatically when the program is loaded for running. The same is true of the stack segment if the program has such a section. The treatment of the data and extra segments is different. Specific instructions to change DS and ES (if used) have to be included in the program. So when the computer is to use the data segment CHAFF, instructions such as

```
mov  ax,chaff
mov  ds,ax
```

will appear in the program. Probably you will have an earlier instruction to save DS for subsequent recovery when your program is finished. Note that when the name of a segment occurs in a MOV it is the value of the segment part of the address which is transferred.

Index-addressing by means of [ ] has been confined to the base registers BX, BP and the index registers SI, DI. Also, the content of the bracket has been limited to one register as in [BX], [DI] though one base register and one index is permitted as in [BP + SI].

Another assembler may offer the facility of declaring *procedures* by means of the directive PROC. The assembler will understand PROC to signal the start of an independent procedure. The procedure must be named and it must be closed with an ENDP directive having a name which matches that of PROC. For instance

```
apple    proc
         ...
apple    endp
```

The PROC can be followed by NEAR or FAR; in the absence of either NEAR will be assumed. NEAR informs the assembler that the subroutine should be assembled close enough to the calling program for a change of code register to be superfluous. Declaring procedures, as opposed to simply calling the label of a set of instructions, tells the assembler about the distance needed to determine how to assemble the RET instruction in the subroutine. When a FAR call is issued (see, for example, CALLF in Appendix B) the code segment is changed. Hence, in order to get back from the call, CS must be preserved. So, before placing the instruction pointer IP on the stack, as described in Section 4.2, CALLF transfers CS to the stack. Therefore, on return, an extra stack operation is involved to recover CS and get the calling program going again at the right spot. The additional stack transfer is attended to automatically if RETF terminates the program that has been called. If CALLF and RETF do not exist on your assembler, and you need far calls you should consult the manual; you may be able to employ RETn.

Some assemblers have the directive LENGTH, which calculates the number of bytes in a variable. Thus

```
buf     db      'ABCDE'
        mov     ax,length buf
```

would result in 5 being placed in AX. If LENGTH is not available the same effect can be achieved by the *location counter* $ (to be distinguished from the string terminator '$'). In essence, when $ appears, it gives the current value of the location counter i.e. the number of bytes in the object code from the beginning. Consequently,

```
buf         db      'ABCDE'
buf_len     equ     $-buf
            mov     ax,buf_len
```

will end up with 5 in AX. The reason is that, if BUF starts 9 bytes from the beginning (recorded as 0009H by the location counter), at the end of the assembler dealing with the DB instruction the location counter will be at 000EH. Because EQU takes no space in the object code the assembler gives $ in BUF_LEN the value 000EH; it then subtracts BUF or 0009H with the result that BUF_LEN is given the value 5.

Another directive which some assemblers offer is INCLUDE. This enables the assembler to incorporate existing programs in your program by assembling them as part of your source code. All that you have to do is to specify the filename (including the drive designation if the file is not on the default drive) and ensure that the file itself does not contain any INCLUDE directives, because they cannot be nested. For example

```
include A:your.lib
```

YOUR.LIB might consist of a series of EQUs which you use frequently and keep on a disk file, though you must remember to have the INCLUDE in your program before you use any of the definitions in your library.

INDEX

AAA 97, 98, 183
AAD 99, 183
AAM 98, 99, 184
AAS 97, 184
absolute value 47
accumulator 24
accumulator, floating point 101
ACK 134, 179
acknowledgement 111
acoustic coupler 122
ADA 16
ADC 60, 96, 97, 98, 184
ADD 28, 184
addition 27, 47, 59
addition in floating point 104
addition of packed BCD 95
addition of unpacked BCD 96
address 2
address value 46
address
  base 13
  direct 12
  direct register 12
  effective 12, 46
  immediate 11
  index 12
  indirect 12, 46, 50
  indirect register 12
  memory 46
  relative 13
ALGOL 16
AND 90, 94, 184
arithmetic 7
arithmetic shift 8
array 44
array
  double 50
  maximum of 47
ASCII 151
  and packed BCD 93
  codes 179–180
  converted to binary 64
  in 7 bits 124
  input 117
  conversion to 61
ASCIZ string 16, 158
assembler 17
  directive 17
  directives 193
  printout 168
ASSUME 194
asynchronous event 109
  signal 125
attribute 45
autovector 112

base addressing 13
BASIC 16, 151
  string parameter 167
  call from 161–7
Basic Disk Operating System 138
Basic Input/Output System 138
baud 123
baud clock 126
BCD 93
  addition of 95, 97
  packed 93, 95
  packed and ASCII 93
  unpacked 93, 96–9
  unpacked division 98
  unpacked multiplication 98
BDOS 138
binary floating point 99
  range 100
  standard format 100
binary search 108
binary, conversion to 64
binary-coded decimal 21, 93
  packed 93
  unpacked 93
BIOS 138
bisection method 108
bit pattern 90
bit, most significant 124
bit, sign 9, 35
bits, concatenation of 90
board game 69
bottom-up 58
branch 10
break signal 128, 133
bubblesort 49, 175
buffer 109, 153
  for communication 123
  video 163
bus 14
bus interface unit 20
byte 15
BYTE PTR 29
bytes, interchange of 91

calendar 75
  updating 117
CALL 60, 184
CALL in BASIC 167

call 11
  from BASIC 161–7
  nested 68
  recursive 68
CALLF 184, 194
carriage return 54, 76
CBW 36, 184
character string 62
CLC 91, 97, 184
CLD 78, 84, 184
clear to send 122
clearing 36
CLI 91, 112, 184
clock updating 117
clock, baud 126
clock, of 8086 21
CMC 91, 184
CMP 39, 184
CMPS 88, 185
CMPSB 87, 185
CMPSW 88, 185
COBOL 19
code segment 25
code
  object 18, 168
  parity-check 137
  re-entrant 56
  source 18, 168
  status 26
codes, ASCII 179–80
combining strings 85
comment 27
comparison 11, 39
compiler 16
concatenation 85
concatenation of bits 90
condition code 26
control key 16
control, transfer of 10
CONTROL-C 139–43
conversion to ASCII 61
conversion to binary 64
copy string 80
counter
  location 195
  program 3
counting 43
CP/M86 177
CSEG 51, 53, 79, 115–18, 130–5, 193
CTS 122
cycle 43
  endless 114
  execute 4
  fetch 4
CWD 36, 185

DAA 95, 185
DAS 96, 185
data carrier detect 122
data declaration 45, 61, 62
data segment 25
data set ready 122
data terminal ready 122
data
  packed 92
  transfer of 9
  unpacked 92
DB 45, 62
DCD 122
DD 62
DEC 43–4, 185
destination 27
device driver 14
Direct Memory Access 109
direction flag 77, 91
directives, assembler 17, 193
disassembly 169
Disk Transfer Address 149
disk drive 2
disk transfers 148–50
disk
  sequential read 149
  sequential write 149
display character 139
display string 140
DIV 36, 185
divide overflow 37
division 32, 34, 36
division of unpacked BCD 98
DMA 109, 115
DMA controller 109
doubleword 21, 60
driver, device 14
DSEG 51, 53, 79, 116–18, 130–5, 193
DSR 122
DTA 149
DTA, setting 149
DTR 122
dummy 76
DW 31, 47, 62

EAROM 13
echo 141
echo of keyboard 141
EDLIN 159
END 52, 53, 154
end process 150
end-of-file character 53
end-of-file, finding 53
end-of-text 135
endless cycle 114
ENDP 194
ENDS 193
EPROM 13

EQU 29–30, 195
error
  computational 100
  framing 126
  representational 100
ESEG 53, 193
ETX 135, 179
executable file 169
execute cycle 4
execution unit 20
exponent 99
  overflow 101
  underflow 101
  offset 100–7
extra segment 25
EXTRN 94, 116, 130, 171, 172

FAR 194
FAT 146
FCB 146–8
fetch cycle 4
File Allocation Table 146
File Control Block 146, 154
file handle 157
file, close 147, 150
  create 147
  executable 169
  length of 163
  open 146
  read from 156
  run 169
  write to 155
files 145–50, 157–61
flag 3
  clearing 3
  lowering 3
  raising 3
  register 3, 91
  setting 3
  auxiliary carry 26, 91, 95–9
  carry 26, 37, 39, 91, 96–9, 107
  direction 26, 77, 91
  interrupt enable 26, 91, 112–14
  load 173
  overflow 26, 37, 39
  parity 26, 91
  sign 26, 37, 91
  trap 26
  zero 26, 38, 91
flags register 25
Floating Point Accumulator 101
Floating Point Register 101
floating point 100
  addition 104
  multiplication 102
FORTRAN 16

frame ground 122
framing error 126

game, board 69
GND 122
ground 122

handle
  close 160
  create 158
  file 157
  open 159
  special input/output 158
handshaking 111, 123, 124
Hanoi, towers of 75
hexadecimal notation 15
HLT 114, 185

iAPX 86/10 20
iAPX 86 177
IBM Macro Assembler 177
IDIV 33, 36, 185
immediate value 11, 28
IMUL 32, 35, 185
IN 77, 116, 186
INC 40, 186
INCLUDE 195
index addressing 12, 194
indirect address 61
input 2, 14, 77
  from keyboard 141
  of character 131
  buffered 142
  standard 155
input/output 115
instruction 5
instruction pointer 3
instruction set of 8086 183
instruction, format of 26
  logical 90
  shift 90
INT 139, 186
INT 21H 139
  function 01H 141
  function 02H 139
  function 05H 140
  function 08H 141
  function 09H 140
  function 0AH 142
  function 0FH 146
  function 10H 147
  function 14H 149
  function 15H 149
  function 16H 147
  function 1AH 149

INT 21H (*cont.*)
  function 25H 162
  function 31H 163
  function 35H 145
  function 3CH 158
  function 3DH 159
  function 3EH 160
  function 3FH 156
  function 40H 155
  function 42H 162
  function 4CH 150
INT 224 139
INTA 113
Intel 8086 microprocessor 20
interface control 128–31
  setting up 130
interrupt 11
  acknowledge 113
  disabling 112
  masking and segment register 112
  request 112
  service routine 11, 111–19
  vector 111, 145
  internal 109
  masked 111
  nonmaskable 111
  priority of 111
  software 140
interrupts 109–19
INTR 112, 130
IP 60
IRET 113–19, 161, 186

J. 192
JA 103, 192
JB 65, 192
JBE 102, 192
JC 39, 105, 192
JCXZ 51, 53, 87, 94, 192
JE 39, 192
JG 40, 81, 192
JGE 81, 192
JMP 37–8, 186
JMPF 38, 186
JMPS 38, 186
JNA 84, 192
JNC 55, 97, 192
JNL 40, 192
JNS 110, 192
JNZ 38, 192
JO 39, 192
JPO 117, 192
JS 110, 116, 192
jump 10
  absolute 10
  conditional 10, 38
  far 38
  unconditional 37
  short 38
JZ 38, 192

keyboard 2
  input 141
kilobyte 15

label 39
  out of range 38
LAHF 91, 186
language
  assembly 17
  compiled 18
  high-level 16
  interpreted 18
  level of 16
LDS 111, 187
LEA 46–54, 94, 97, 187
left of string 81
LENGTH 195
length of string 78
LES 111, 187
level of language 16
level
  mark 125
  space 125
line feed 54, 76
link on stack 66
link register 66
linker 18, 169
load flag 173
location counter 195
logic 7
  shift 8
  p AND q 7
  p OR q 7, 8
  p XOR q 7, 8
logical instruction 90
loop 43
LOOP 65, 97, 187
LOOPE 65, 187
LOOPNE 65, 77, 187
LOOPNZ 65, 187
LOOPZ 65, 187

machine language 168
macro 143–5
magnitude, unsigned 33
manager, overlay 172
mark level 125
mask 7, 95
maximum mode 20
megabyte 15

memory 2
  random access 13
  read-only 13
  segment 21
menu 170
microprocessor 2
middle of string 83
minimum mode 20
mnemonic 17, 27
mnemonics for 8086 183
modem 2, 122
module 58
modulus 47
monitor 2
Motorola 68000 20
MOV 29, 166, 188, 194
MOVS 80, 188
MOVSB 80, 83, 84, 188
MOVSW 80, 188
MS-DOS 177
MSB 124
MUL 35, 98, 188
multiple instructions 116
multiplication 32, 34
  in floating point 102
  of unpacked BCD 98

NAK 134, 179
NEAR 194
NEG 48, 49, 105, 188
NMI 111
no operation 11
NOP 11, 188
NOT 90, 188
NOT p 7
number
  binary 15
  hexadecimal 15
  octal 15
  signed 34

object code 18, 168
offset 21
OFFSET 141
offset exponent 100–7
opcode 27
operand 27
  dummy 76
OR 90, 94, 103, 188
ORG 56, 115, 117
OUT 117, 188
output 2, 14
  of character 131
  standard 155
overflow 35
  divide 37
  in exponent 101
overlay 170–3
overlay manager 172
overlays, nesting of 173
override prefix 77

packing data 92
packing, Intel convention for 95
page 15
PARA 166
paragraph 15, 163
parallel transmission 121
parameter 58
parity 124
parity-check code 137
PASCAL 16, 58, 100
password 141
PATH 168
pathname 158
PEEK 167
pointer 29, 46
  instruction 3, 25
  read/write 159
  stack 4
POKE 167
polling 109
POP 4, 5, 63, 64, 188
POPF 92, 188
port 2, 115, 130–5
postdecrement 13
postincrement 12
precision of representation 99
predecrement 13
prefix, override 77
preincrement 12
print character 140
printer 2
PROC 194
procedure 194
process, end 150
processor, central 2
program 4
Program Segment 146
Program Segment Prefix 146, 154
program counter 3
  exit 150
  fragment 28
  relocatable 25
programming, structured 173–4
  techniques for 19
protocol 122
  for interface 127
pseudo-op 17
PTR 29, 45, 46, 119
PULL 4
PUSH 4, 5, 63, 64, 188
PUSHF 92, 188

radix 15
RAM 13
RAM disk 138
random access memory 13
RB 76
read keyboard 141
read keyboard and echo 141
read-only memory 13
read/write pointer 159
  move 162
receive data 122
receiver buffer overflow 129
receiver overrun 129
record 148
  partial 148
recursion 68, 73–4
redirection 155, 157
refinement 19
register 3
  address 3
  base 61
  clearing of 36
  data 3
  flags 3, 25, 91
  floating point 101
  index 3, 61
  instruction 3
  memory address 3
  segment 25
  status 3
registers of 8086 21
REP 80, 83, 84, 189
REPE 81, 87, 189
REPNE 81, 87, 189
request to send 122
RESET 112
RET 60, 189
RET n 68, 166, 189, 195
RETF 166, 189, 195
return instruction 11
right of string 82
ROL 91, 103, 189
ROM 13
ROR 91, 189
rotation 8, 91
rounding 100
routine, library 169
RS 31, 45
RTS 122
run file 169
RW 45
RXD 122

SAHF 92, 189
SAL 8, 190
SAR 8, 190
scale factor 99
SCAS 79, 190
SCASB 79, 87, 190
SCASW 79, 190
SEG 162
SEGMENT 193
segment override 77
  register 25
  register and interrupt 112
segment, memory 21
serial transmission 121
shell sort 175
shift 8
  instruction 90
  arithmetic 8
  logic 8, 90
  rotate 8
SHL 94, 103, 106, 190
SHR 105, 106, 190
sign bit 9, 35
signal, asynchronous 125
signal, synchronous 125
SI/SO 124–5, 179
software interrupt 140
source 27
source code 18, 168
space level 125
SSEG 53, 193
stack 4
  growth 66
  linking 66
  pointer 4
  segment 25
standard input 155
standard output 155
start bit 126–31
status code 26
STC 91, 190
STD 83, 190
STDIN 155
STDOUT 155
STI 91, 112, 190
stop bit 126–31
STOS 76–8, 190
STOSB 78, 190
STOSW 78, 190
string address increment 77
string descriptor 167
  destination address 76
  parameter in BASIC 167
  source address 76
  character 62
  copy of 80
  left of 81
  length of 78
  middle of 83
  right of 82
  storage of 77
  substring within 86

strings, concatenation of 85
SUB 31, 191
subroutine 11
substring 81
  location of 86
subtraction 31
synchronous signal 125
syntax errors 169
system calls 138

terminate, leaving resident 163
TEST 103, 105, 106, 110, 191
top-down 58
towers of Hanoi 75
transmission, 7-bit 124
transmission, parallel 14, 121
  serial 14, 121
transmit data 122
truncation 100
TXD 122

unassembly 169
underflow in exponent 101, 107
underscore 27
unit, central processing 1
unpacked data 92
updating clock 117
use of ! 116

value, immediate 11
variable, local 59
vector address, setting 162
vector table 111, 161
vector, interrupt 111
video buffer 163

word 15
WORD PTR 29, 40

XCHG 91, 191
XON/XOFF 123, 179
XOR 36, 39, 90, 191

zero flag 10